Elisabeth Rakus-Andersson, Ronald R. Yager, Nikhil Ichalkaranje,
and Lakhmi C. Jain (Eds.)

Recent Advances in Decision Making

Studies in Computational Intelligence, Volume 222

**Editor-in-Chief**
Prof. Janusz Kacprzyk
Systems Research Institute
Polish Academy of Sciences
ul. Newelska 6
01-447 Warsaw
Poland
*E-mail:* kacprzyk@ibspan.waw.pl

Further volumes of this series can be found on our homepage: springer.com

Vol. 201. Aboul-Ella Hassanien, Ajith Abraham, Athanasios V. Vasilakos, and Witold Pedrycz (Eds.)
*Foundations of Computational Intelligence Volume 1,* 2009
ISBN 978-3-642-01081-1

Vol. 202. Aboul-Ella Hassanien, Ajith Abraham, and Francisco Herrera (Eds.)
*Foundations of Computational Intelligence Volume 2,* 2009
ISBN 978-3-642-01532-8

Vol. 203. Ajith Abraham, Aboul-Ella Hassanien, Patrick Siarry, and Andries Engelbrecht (Eds.)
*Foundations of Computational Intelligence Volume 3,* 2009
ISBN 978-3-642-01084-2

Vol. 204. Ajith Abraham, Aboul-Ella Hassanien, and André Ponce de Leon F. de Carvalho (Eds.)
*Foundations of Computational Intelligence Volume 4,* 2009
ISBN 978-3-642-01087-3

Vol. 205. Ajith Abraham, Aboul-Ella Hassanien, and Václav Snášel (Eds.)
*Foundations of Computational Intelligence Volume 5,* 2009
ISBN 978-3-642-01535-9

Vol. 206. Ajith Abraham, Aboul-Ella Hassanien, André Ponce de Leon F. de Carvalho, and Václav Snášel (Eds.)
*Foundations of Computational Intelligence Volume 6,* 2009
ISBN 978-3-642-01090-3

Vol. 207. Santo Fortunato, Giuseppe Mangioni, Ronaldo Menezes, and Vincenzo Nicosia (Eds.)
*Complex Networks,* 2009
ISBN 978-3-642-01205-1

Vol. 208. Roger Lee, Gongzu Hu, and Huaikou Miao (Eds.)
*Computer and Information Science 2009,* 2009
ISBN 978-3-642-01208-2

Vol. 209. Roger Lee and Naohiro Ishii (Eds.)
*Software Engineering, Artificial Intelligence, Networking and Parallel/Distributed Computing,* 2009
ISBN 978-3-642-01202-0

Vol. 210. Andrew Lewis, Sanaz Mostaghim, and Marcus Randall (Eds.)
*Biologically-Inspired Optimisation Methods,* 2009
ISBN 978-3-642-01261-7

Vol. 211. Godfrey C. Onwubolu (Ed.)
*Hybrid Self-Organizing Modeling Systems,* 2009
ISBN 978-3-642-01529-8

Vol. 212. Viktor M. Kureychik, Sergey P. Malyukov, Vladimir V. Kureychik, and Alexander S. Malyoukov
*Genetic Algorithms for Applied CAD Problems,* 2009
ISBN 978-3-540-85280-3

Vol. 213. Stefano Cagnoni (Ed.)
*Evolutionary Image Analysis and Signal Processing,* 2009
ISBN 978-3-642-01635-6

Vol. 214. Been-Chian Chien and Tzung-Pei Hong (Eds.)
*Opportunities and Challenges for Next-Generation Applied Intelligence,* 2009
ISBN 978-3-540-92813-3

Vol. 215. Habib M. Ammari
*Opportunities and Challenges of Connected k-Covered Wireless Sensor Networks,* 2009
ISBN 978-3-642-01876-3

Vol. 216. Matthew Taylor
*Transfer in Reinforcement Learning Domains,* 2009
ISBN 978-3-642-01881-7

Vol. 217. Horia-Nicolai Teodorescu, Junzo Watada, and Lakhmi C. Jain (Eds.)
*Intelligent Systems and Technologies,* 2009
ISBN 978-3-642-01884-8

Vol. 218. Maria do Carmo Nicoletti and Lakhmi C. Jain (Eds.)
*Computational Intelligence Techniques for Bioprocess Modelling, Supervision and Control,* 2009
ISBN 978-3-642-01887-9

Vol. 219. Maja Hadzic, Elizabeth Chang, Pornpit Wongthongtham, and Tharam Dillon
*Ontology-Based Multi-Agent Systems,* 2009
ISBN 978-3-642-01903-6

Vol. 220. Bettina Berendt, Dunja Mladenic, Marco de de Gemmis, Giovanni Semeraro, Myra Spiliopoulou, Gerd Stumme, Vojtech Svatek, and Filip Zelezny (Eds.)
*Knowledge Discovery Enhanced with Semantic and Social Information,* 2009
ISBN 978-3-642-01890-9

Vol. 221. Tassilo Pellegrini, Sören Auer, Klaus Tochtermann, and Sebastian Schaffert (Eds.)
*Networked Knowledge - Networked Media,* 2009
ISBN 978-3-642-02183-1

Vol. 222. Elisabeth Rakus-Andersson, Ronald R. Yager, Nikhil Ichalkaranje, and Lakhmi C. Jain (Eds.)
*Recent Advances in Decision Making,* 2009
ISBN 978-3-642-02186-2

Elisabeth Rakus-Andersson, Ronald R. Yager,
Nikhil Ichalkaranje, and Lakhmi C. Jain (Eds.)

# Recent Advances in Decision Making

Springer

Prof. Elisabeth Rakus-Andersson
Blekinge Institute of Technology
School of Engineering
Department of Mathematics and Science
S-37179 Karlskrona
Sweden
E-mail: era@bth.se

Ronald R. Yager
Iona College
Machine Intelligence Institute
715 N. Avenue
New Rochelle NY 10801-1890
USA
E-mail: yager@panix.com

Dr. Nikhil Ichalkaranje
School of Electrical and
Information Engineering
University of South Australia
South Australia SA 5095
Australia

Prof. Lakhmi C. Jain
School of Electrical and Information
Engineering
University of South Australia
Adelaide
Mawson Lakes Campus
South Australia SA 5095
Australia
E-mail: Lakhmi.jain@unisa.edu.au

ISBN 978-3-642-02186-2 e-ISBN 978-3-642-02187-9

DOI 10.1007/978-3-642-02187-9

Studies in Computational Intelligence ISSN 1860949X

Library of Congress Control Number: Applied for



*Typeset & Cover Design:* Scientific Publishing Services Pvt. Ltd., Chennai, India.

Printed in acid-free paper

9 8 7 6 5 4 3 2 1

springer.com

# Preface

It is not only the vast amount of data but knowledge extraction and processing play an important role in the design of the decision support systems. Sensible decision support systems are required in virtually every field including business, healthcare, defence and so on [1]. The design of decision support systems is also dependent on factors such as changing sets of circumstances, uncertainty, incomplete set of data. Intelligent paradigms such as knowledge-based systems, artificial neural networks, fuzzy systems, evolutionary computing paradigms, intelligent agents have contributed immensely in the decision making process. Knowledge-based systems [2] can mimic the performance of a human expert in a limited sense by transferring his/her knowledge to the computer in a specific domain. Artificial neural networks are modeled after the human brain for fusing human like intelligence in machines. Fuzzy systems are designed to incorporate human like reasoning capability in machines. Evolutionary systems use principles inspired by natural population genetics and are applied in many problems including optimization. Intelligent agents can aid and automate complex problem solving in many areas and help in effective decisions [3]. The combination of intelligent systems and decision support systems provides new powerful tools for decision makers [4][5].

The book is a collection of selected contributions from some of the world class researchers in the field of intelligent tools and decision making. This sample is to demonstrate that the intelligent tools can enhance the decision making process.

We sincerely thank the contributors and reviewers for their excellent contribution. We acknowledge the excellent support of Springer-Verlag and SCI Data Processing Team.

Elisabeth Rakus-Andersson, Sweden
Ronald R. Yager, USA
Nikhil Ichalkaraje, Australia
Lakhmi C. Jain, Australia

## References

[1] Tonfoni, G., Jain, L.C.: Innovations in Decision Support Systems, Advanced Knowledge International (2003)
[2] Jain, L.C., Jain, R.K.: Hybrid Intelligent Engineering systems. World Scientific, Singapore (1997)
[3] Phillips-Wren, G.E., Jain, L.C.: Intelligent Decision Support Systems in Agent Mediated-Environments. IOS Press, Amsterdam (2005)
[4] Phillips-Wren, G.E., Ichalkaranje, N., Jain, L.C.: Intelligent Decision Making: An AI-Based Approach. Springer, Heidelberg (2008)
[5] Teodorescu, H.-N., Kandel, A., Jain, L.C.: Fuzzy and Neuro-fuzzy Systems in Medicine. CRC Press, USA (1999)

# Editors

Professor Dr Elisabeth Rakus-Andersson is employed at Blekinge Institute of Technology in Karlskrona, Sweden. In her research she focuses on imprecise mathematical theories like fuzzy set theory or rough set theory. As she has a background in mathematics, she develops own models mathematically formalized and she even adapts patterns already existing to generate new applications in medicine and technical environments.

Ronald R. Yager is Director of the Machine Intelligence Institute and Professor of Information Systems at Iona College. He is editor and chief of the International Journal of Intelligent Systems. He has worked in the area of machine intelligence and decision making under uncertainty for over twenty-five years. He has published over 500 papers and fifteen books. He is among the world's top 1% most highly cited researchers with over 7000 citations. He was the recipient of the IEEE Computational Intelligence Society Pioneer award in Fuzzy Systems. Dr. Yager is a fellow of the IEEE, the New York Academy of Sciences and the Fuzzy Systems Association. He was given a lifetime achievement award by the Polish Academy of Sciences. He served at the National Science Foundation as program director in the Information Sciences program. He was a NASA/Stanford visiting fellow and a research associate at the University of California, Berkeley. He has been a lecturer at NATO Advanced Study Institutes. He has been a distinguished honorary professor at the Aalborg University Esbjerg Denmark. He is an affiliated distinguished researcher at the European Centre for Soft Computing. He serves on the editorial board of numerous technology journals.

Nikhil Shripal Ichalkaranje is Senior Technology Advisor in the Australian Department of Broadband, Communications and the Digital Economy. He is also an adjunct Senior Research Fellow at the University of South Australia. His research interests include artificial intelligence, computer communications/networking and robotics. Nikhil holds a Masters Degree in Computer Network Systems from Swinburne University of Technology Melbourne and a PhD in Computer Systems Engineering from the University of South Australia Adelaide. Nikhil has co-edited 5 books along with several publications in the area of artificial intelligence and their applications.

Professor L.C. Jain is a Director/Founder of the Knowledge-Based Intelligent Engineering Systems (KES) Centre, located in the University of South Australia.

His interests focus on the applications of novel techniques such as knowledge-based systems, virtual intelligent systems, defence systems, intelligence-based medical systems, e-Education and intelligent agents.

# Table of Contents

1

# Advances in Decision Making

Lakhmi C. Jain[1] and Chee Peng Lim[2]

[1] School of Electrical & Information Engineering
University of South Australia, Australia
[2] School of Electrical & Electronic Engineering
University of Science Malaysia, Malaysia

**Abstract.** This chapter presents the application of Computational Intelligence (CI) paradigms for supporting decision making processes. First, the three main CI techniques, i.e., evolutionary computing, fuzzy computing, and neural computing, are introduced. Then, a review of recent applications of CI-based systems for decision making in various domains is presented. The contribution of each chapter included in this book is also described. A summary of concluding remarks is presented at the end of the chapter.

## 1 Introduction

The task of decision making occurs in our daily life. Decision making arises from the need to select the best possible course of action (or a set of optimized actions) from a set of alternatives. We are presented with a lot of information and/or data in our daily activities, and we, either consciously or sub-consciously, have to make decisions based on the received information and/or data. However, making a good and accurate decision is a challenging task. This is because conflicts and tradeoffs often surface owing to the multiple objectives and goals that are to be simultaneously satisfied by the decision maker.

The demand for prompt and accurate decision making is exacerbated by the rapid development and wide-spread usage of the internet as a resource for information and knowledge sharing and reuse. Indeed, the world-wide-web contains many heterogeneous information sources ranging from text documents to multimedia images; from audio files to video streams. Thus, the process of information and data generation and acquisition has become easy and almost instantaneous. Nevertheless, the information and/or data gathered from real-world systems or processes often are complex and multi-facet, and comprise various kinds of noise. Besides, real world systems or processes often produce incomplete information and/or data owing to the unavailability of system parameters or structures, as well as the uncertainties of the environment in which the system or process operates.

To cope with the challenges of decision making, researchers have investigated and proposed a variety Decision Support Systems (DSSs) to provide assistance in the process of decision making. In general, a DSS is a computerized information system that supports decision-making activities in various domains such as business, finance, management, manufacturing, and biomedicine. A useful conceptual framework for DSS classification is proposed by Power (2004) [10]. Five generic types of DSSs are

E. Rakus-Andersson et al. (Eds.): Recent Advances in Decision Making, SCI 222, pp. 1–6.
springerlink.com 

identified and defined based on the dominant technology component. These are communications-driven, data-driven, document-driven, knowledge-driven, and model-driven. A DSS can be developed for either specific or general-purpose applications, and can be used by either individuals or groups. The enabling technology of the DSS can be a mainframe computer, a client/server LAN, a spreadsheet, or a web-based architecture (Power, 2004) [10].

In a technology field as diverse as DSS, many methodologies have been proposed to help build and understand these systems. Nevertheless, the adage that "all roads lead to Rome" applies, i.e., the ultimate goal of these methods and techniques for tackling decision making problems is to help humans make informed decisions timely and accurately. One of research trends is geared towards designing a new generation of intelligent DSSs that possess a high level of machine learning quotient. In this chapter, we introduce intelligent computing paradigms under the umbrella of Computational Intelligence (CI) for designing and developing intelligent DSSs. Note that only a small fraction of DSSs that utilize CI-based technologies for decision making is introduced in this chapter. The main aim is to share and disseminate information pertaining to recent advancements in developing intelligent DSSs for tackling real-life problems in various domains.

The organisation of this chapter is as follows. In section 2, an introduction to the main characteristics of three main CI paradigms is presented. Applicability of various CI-based as well as hybrid CI-based systems to tackling decision making in different domains is reviewed. The contribution of each chapter in this book is described in section 3. Section 4 gives some concluding remark of this chapter.

## 2 Computational Intelligence-Based Decision Making Systems

CI is an interdisciplinary field that is useful for supporting the design and development of intelligent systems. According to Bezdek (1994) [3], "… a system is computationally intelligent when it: deals only with numerical (low-level) data, has a pattern recognition component, and does not use knowledge in the AI sense; and additionally when it (begins to) exhibit (i) computational adaptivity; (ii) computational fault tolerance; (iii) speed approaching human-like turnaround, and (iv) error rates that approximate human performance …". Marks (1993) [8] explained that "… neural networks, genetic algorithms, fuzzy systems, evolutionary programming, and artificial life are the building blocks of CI …".

While a variety of computing paradigms are available nowadays, we focus on three main CI constituents, i.e. evolutionary computing (EC), fuzzy computing (FC), and neural computing (NC), as well as their hybrid models. In particular, we first present an overview of EC, FC, and NC that provide a platform for intelligent decision making. Then, we review the recent applications of CI-based systems for tackling real-world decision making problems in a variety of domains.

### 2.1 Overview of the Three Main CI-Based Systems

EC-based systems operate on the principles of evolution and natural selection in living population. They are population-based algorithms for which any communication and interaction are carried out within the population. On availability of a population of individuals, natural selection, which is based on the principle of survival of the fittest following the existence of environmental pressures, is exercised to choose individuals that could better fit the environment. EC-based systems normally possess a high degree of implicit parallelism, and are particularly useful for applications that require search and optimization.

FC-based systems assimilate the concepts of fuzzy set theory and fuzzy logic that provide a framework for handling commonsense knowledge represented in a linguistic or an uncertain numerical form. They are useful for representing and reasoning with uncertain, imprecise, and vague data and/or information. Fuzzy logic provides an inference mechanism on a set of *if-then* rules for reasoning. The rules are defined with fuzzy sets, in which the fuzzy sets generalize the concept of the conventional set by extending membership degree to be any value between 0 and 1. Such "fuzziness" feature occurs in many real-world situations, where it is ambiguous to decide if something can be categorized exactly into a specific class or not.

NC-based systems attempt to mimic certain functions of the brain at the macroscopic level such as capturing and processing information. From the machine learning point of view, NC-based systems are capable of forming a non-linear mapping between a set of input-output data samples. They can be employed as universal functional approximators, which can offer accurate approximation of an unknown model on provision of data samples. Some of the NC-based systems are enhanced with the capability of absorbing information continually and autonomously without forgetting previously learned information. Such ability is favourable for the systems to operate in non-stationery environments with greater autonomy and less dependency on humans.

While EC, FC, and NC-based systems can be applied independently to solve real-world problems, more effective solutions can be obtained if they are used in combination. Examples of hybrid CI-based systems include neural-fuzzy, neural-genetic, fuzzy-genetic, and neural-fuzzy-genetic models. In addition, other machine learning techniques can also be integrated with these CI constituents such as rule-based systems, decision trees, case-based reasoning, knowledge-based systems, probabilistic reasoning and rough sets. Indeed, hybrid CI-based system systems are increasingly popular owing to the synergy that exploits the advantages of each intelligent technique and, at the same time, avoids its shortcomings.

## 2.2 Application Examples of CI-Based Systems to Decision Making

In civil and structural engineering, EC, in particular the genetic algorithm (GA), has been applied to support decision making in conceptual building designs (Rafiq *et. al.*, 2005) [11]. In conceptual building design support, knowledge-based systems face two difficulties: elucidation of expert knowledge and the nature of the crisp rules. The GA, which does not rely on a predefined rule set and which is able to evolve the solutions, appear to be a strong candidate. The GA, coupled with interactive visualization and clustering techniques, has been found to be useful in providing alternative solutions that could be assessed by building designers against a set of predefined requirements (Rafiq *et. al.*, 2005) [11].

A GA-based decision support system has been built for determining the optimal budget allocation and relevant contracting methods in historical building restoration and preservation in Taiwan (Perng *et. al.*, 2007) [9]. The system is able to provide more effective and economical decision suggestions as compared with the traditional contracting methods.

A hybrid approach is applied to tackle industrial design of a racing car tire-suspension system (Benedetti *et. al.*, 2007) [2]. The problem is complex as it involves 24 objective functions, with 18 conflicting with each other. The proposed hybrid approach, which involves evolutionary multi-objective optimization, neural network modelling, as well as fuzzy optimality-based analysis, is useful for supporting this challenging decision making problem.

Multiple criteria decision making is widely used in ranking one or more alternatives from a set of available alternatives with respect to multiple criteria. In Cebeci (2009) [4], the fuzzy AHP (Analytic Hierarchy Process), which is a fuzzy extension of the multi-criteria decision-making technique of AHP, is used to select a suitable ERP (Enterprise Resource Planning) system for a textile manufacturing company. On the other hand, a fuzzy TOPSIS (Technique for Order Preference by Similarity to Ideal Solution) approach is employed in Wang & Lee (2009) [14] for evaluating alternatives by integrating not only subjective weights from decision makers but also end-user ratings as an objective weight based on Shannon's entropy theory. An interesting model that combines AHP and TOPSIS to support decision making in weapon selection in a fuzzy environment is described in Dağdeviren *et. al.* (2009) [5]. The vagueness and subjectivity are handled with linguistic values parameterized by triangular fuzzy numbers. The structure of the weapon selection problem and the weights of the criteria are analysed using AHP, while the final ranking is determined based on the fuzzy TOPSIS method.

Conflicts always occur in group decision making as group members generally do not have a consensus on a specific issue. In Wu (2009) [15], a fuzzy group decision making setting with grey related analysis and Dempster–Shafer theory is described. Grey related analysis is employed as a means to reflect uncertainty in multi-attribute models through interval numbers, while the Dempster–Shafer rule of combination is used to aggregate individual preferences into a collective preference. The applicability of the proposed approach to international supplier selection is demonstrated.

In medical applications, a new tremor diagnosis approach based on multi-features extraction, the back-propagation neural network, and the Dempster-Shafer evidence theory is proposed (Ai *et. al.*, 2008) [1]. The proposed system is able to utilize the complementary multi-features information for accurately recognizing different types of tremor, thus providing decision support for diagnosing tremor types in clinical practice. The ANFIS (adaptive neuro-fuzzy inference system) model is used for classification of electrocardiogram (ECG) signals (Übeyli, 2009a) [12]. The process of decision making is accomplished in two stages: feature extraction by computation of Lyapunov exponents and classification by the ANFIS. This approach essentially combines the neural network adaptive capabilities and the fuzzy logic qualitative capabilities for decision making. A recurrent neural network combined with the eigenvector method is also used to tackle a similar problem (Übeyli, 2009b) [13].

In Jarman et. al. (2008) [7], the orthogonal search rule extraction method is used to generate interpretable explanations of risk group allocations derived from a partial logistic artificial neural network with automatic relevance determination (PLANN–ARD) in an attempt to develop an integrated framework for risk profiling of breast cancer patient following surgery. The C4.5 decision tree and the backpropagation neural network are deployed to construct decision support systems for predicting the regimen adequacy of vancomycin, an antibiotic effective for Gram-positive bacterial infections (Hu *et. al.*, 2007) [6]. The results indicate that the C4.5 or neural network-based decision support system performs better than that of the benchmark one-compartment pharmacokinetic model.

## 3 Chapters Included in This Book

This book includes nine chapters on theory and technologies of decision making in intelligent environment. Chapter one introduces the book to the readers. Chapter two is on information and its reliability in the ranking of Atanassov's intuitionistic fuzzy alternatives. Chapter three is on fuzzy rule base model identification by bacterial memetic algorithms. Chapter four is on discovering associations with uncertainty from large databases. These associations can be used to make informed decisions.

Chapter five is on Dempster-Shafer structures, monotonic set measures and decision making. The author has formulated the problem of decision making under uncertainty and presented a generalized framework for decision making. Chapter six presents the development of interpretable decision making models. The author has considered two design problems and shown the solution using fuzzy sets. Chapter seven presents a general methodology for managerial decision making using intelligent techniques. The methodology is demonstrated using an example. Chapter eight is on supporting decision making via verbalization of data analysis results using linguistic data summaries. The technique is validated using an example of computer retailer. Chapter nine is on approximate reasoning in surgical decisions. The approximate reasoning considers evaluation of a risk in the situation when physicians weigh necessity of the operation on a patient. The mathematical model presented in the chapter is applicable to other healthcare related applications.

## 4 Summary

This chapter has presented an overview of CI-based paradigms for handling decision making tasks. Owing to information overload in today's digital era, DSSs are becoming important to help extract and elicit meaningful and actionable information and knowledge for decision makers to make informed decisions. In this aspect, CI paradigms as well as their hybrid systems offer a suitable platform for developing intelligent DSSs in various domains. It is envisaged that hybrid CI systems will eventually gain popularity owing to the synergy that exploits the advantages of each intelligent technique and, at the same time, avoids its shortcomings in tackling complex decision making problems.

## References


[1] Ai, L., Wang, J., Wang, X.: Multi-features fusion diagnosis of tremor based on artificial neural network and D–S evidence theory. Signal Processing 88, 2927–2935 (2008)

[2] Benedetti, A., Farina, M., Gobbi, M.: Evolutionary multiobjective industrial design: the case of a racing car tire-suspension system. IEEE Trans. on Evolutionary Computation 10, 230–244 (2006)

[3] Bezdek, J.C.: What is a computational intelligence? In: Zurada, J.M., Marks II, R.J., Robinson, C.J. (eds.) Computational Intelligence: Imitating Life, pp. 1–12. IEEE Press, Los Alamitos (1994)

[4] Cebeci, U.: Fuzzy AHP-based decision support system for selecting ERP sys-tems in textile industry by using balanced scorecard. Expert Systems with Applications 36, 8900–8909 (2009)

[5] Dağdeviren, M., Yavuz, S., Kılınç, N.: Weapon selection using the AHP and TOPSIS methods under fuzzy environment. Expert Systems with Applications 36, 8143–8151 (2009)

[6] Hu, P.J.H., Wei, C.P., Cheng, T.H., Chen, J.X.: Predicting adequacy of vancomycin regimens: A learning-based classification approach to improving clinical decision making. Decision Support Systems 43, 1226–1241 (2007)

[7] Jarman, I.H., Etchells, T.A., Martin, J.D., Lisboa, P.J.G.: An integrated framework for risk profiling of breast cancer patients following surgery. Artificial Intelligence in Medicine 42, 165–188 (2008)

[8] Marks, R.: Intelligence: computational versus artificial. IEEE Transactions on Neural Networks 4, 737–739 (1993)

[9] Perng, Y.H., Juan, Y.K., Hsu, H.S.: Genetic-algorithm-based decision support for the restoration budget allocation of historical buildings. Building and Environment 42, 770–778 (2007)

[10] Power, D.J.: Specifying an expanded framework for classifying and describing decision support systems. Communications of the Association for Information Systems 13, 158–166 (2004)

[11] Rafiq, Y., Beck, M., Packham, I., Denhan, S.: Evolutionary computation and visualisation as decision support tools for conceptual building design. In: Topping, B.H.V. (ed.) Innovation in Civil and Structural Engineering Computing, pp. 49–74. Saxe-Coburg Publications (2005)

[12] Übeyli, E.D.: Adaptive neuro-fuzzy inference system for classification of ECG signals using Lyapunov exponents. Computer Methods and Programs in Biomedicine 93, 313–321 (2009a)

[13] Übeyli, E.D.: Combining recurrent neural networks with eigenvector methods for classification of ECG beats. Digital Signal Processing 19, 320–329 (2009b)

[14] Wang, T.C., Lee, H.D.: Developing a fuzzy TOPSIS approach based on sub-jective weights and objective weights. Expert Systems with Applications 36, 8980–8985 (2009)

[15] Wu, D.: Supplier selection in a fuzzy group setting: a method using grey related analysis and Dempster–Shafer theory. Expert Systems with Applications 36, 8892–8899 (2009)

2

# Amount of Information and Its Reliability in the Ranking of Atanassov's Intuitionistic Fuzzy Alternatives

Eulalia Szmidt and Janusz Kacprzyk

Systems Research Institute, Polish Academy of Sciences,
ul. Newelska 6, 01–447 Warsaw, Poland
Warsaw School of Information Technology,
ul. Newelska 6, 01-447 Warsaw, Poland
{szmidt,kacprzyk}@ibspan.waw.pl

**Abstract.** In this paper we discuss the ranking of alternatives represented by elements of Atanassov's intuitionistic fuzzy sets, to be called A-IFSs, for short. That is, alternatives are elements of the universe of discourse with a degree of membership and a degree of non-membership assigned. First, we show disadvantages of some approaches known from the literature, including a straightforward method based on the calculation of distances from the ideal positive alternative which can be viewed as a counterpart of the approach in the traditional fuzzy setting. Instead, we propose an approach which takes into account not only the amount of information related to an alternative (expressed by a distance from an ideal positive alternative) but also the reliability of information represented by an alternative meant as how sure the information is.

## 1 Introduction

Atanassov's intuitionistic fuzzy sets (cf. Atanassov [2], [3]), to be called A-IFSs for brevity, which are a generalization of the fuzzy sets(Zadeh [35]) can be viewed as a tool that may help better model imperfect information, especially under imperfectly defined facts and imprecise knowledge. A-IFSs have found numerous applications in many areas, notably decision making. One of important, omnipresent problems in the context of decision making, and many other contexts, is the ranking of fuzzy (or intuitionistic fuzzy) alternatives (options), for instance obtained as a result of decision analysis, evaluation, aggregation, etc. The fuzzy and intuitionistic fuzzy alternatives may be understood in different ways, and in this paper we meant them, in the fuzzy case, as elements of the universe of discourse with their associated membership degrees, and, in the intuitionistic fuzzy case, as elements of a universe of discourse with their associated membership and non-membership degrees. We consider here the latter case, and then a natural interpretation in our context of decision making can be that each option fulfills a set of criteria to some extent $\mu(.)$ and, on the other hand, it does not fulfill this set of criteria to some extent $\nu(.)$. This clearly suggest that the alternatives can conveniently

E. Rakus-Andersson et al. (Eds.): Recent Advances in Decision Making, SCI 222, pp. 7–19.
springerlink.com 

be expressed via Atanassov's intuitionistic fuzzy sets. For brevity, such alternatives will be called *intuitionistic fuzzy alternatives*.

The problem of ranking intuitionistic fuzzy alternatives may be solved under some additional assumptions only because there is no linear order among elements of the A-IFSs as opposed to that for fuzzy sets (Zadeh [35]) for which elements of the universe of discourse are naturally ordered because their membership degrees are real numbers from $[0, 1]$.

In the literature there are not many approaches for ranking the intuitionistic fuzzy alternatives. They were proposed by, for instance, Chen and Tan [5], Hong and Choi [7], Li et al. [8], [9], and Hua-Wen Liu and Guo-Jun Wang [10].

Here we propose another approach that is different in several respects.

First, we employ the representation of A-IFSs, which constitute the representation of intuitionistic fuzzy alternatives, taking into account all three functions: the membership function, non-membership function, and hesitation margin. Such a representation has proved to be effective and efficient in solving many problems giving intuitively appealing results (cf. e.g., Szmidt and Kacprzyk[28], [21], [30]), [31]) while constructing measures of a distance, similarity, entropy, etc. that play a crucial role in virtually all information processing tasks, notably those related to decision making.

Second, we propose an ordering function for ranking intuitionistic fuzzy alternatives which depends on two factors: the amount of information associated with an alternative (expressed by the distance from the ideal positive alternative), and the reliability of information (i.e. how sure an alternative is) – expressed by the hesitation margin.

As an example we present an application to a choice of a best course of action in the context of medical treatment.

## 2 A Brief Introduction to Intuitionistic Fuzzy Sets

One of the possible generalizations of a fuzzy set in $X$ (Zadeh [35]), given by

$$A^{'} = \{< x, \mu_{A^{'}}(x) > | x \in X\} \tag{1}$$

where $\mu_{A^{'}}(x) \in [0, 1]$ is the membership function of the fuzzy set $A^{'}$, is Atanassov's intuitionistic fuzzy set (Atanassov [1], [2], [3]) $A$ given by

$$A = \{< x, \mu_A(x), \nu_A(x) > | x \in X\} \tag{2}$$

where: $\mu_A : X \to [0, 1]$ and $\nu_A : X \to [0, 1]$ such that

$$0 \leq \mu_A(x) + \nu_A(x) \leq 1 \tag{3}$$

and $\mu_A(x)$, $\nu_A(x) \in [0, 1]$ denote the degree of membership and a degree of non-membership of $x \in A$, respectively.

Obviously, each fuzzy set may be represented by the following intuitionistic fuzzy set

$$A = \{< x, \mu_{A^{'}}(x), 1 - \mu_{A^{'}}(x) > | x \in X\} \tag{4}$$

For each intuitionistic fuzzy set in $X$, we will call

$$\pi_A(x) = 1 - \mu_A(x) - \nu_A(x) \tag{5}$$

an *intuitionistic fuzzy index* (or a *hesitation margin*) of $x \in A$ and, it expresses a lack of knowledge of whether $x$ belongs to $A$ or not (cf. Atanassov [3]). It is obvious that $0 \leq \pi_A(x) \leq 1$, for each $x \in X$.

The hesitation margin turns out to be important while considering the distances (Szmidt and Kacprzyk [15], [19], [28], entropy (Szmidt and Kacprzyk [21], [30]), similarity (Szmidt and Kacprzyk [31]) for the A-IFSs, etc. i.e., the measures that play a crucial role in virtually all information processing tasks. In this paper the hesitation margin is shown to be useful, if not indispensable, in ranking the intuitionistic fuzzy alternatives because it indicates how reliable (sure) the information represented by an alternative is.

The use of A-IFSs instead of fuzzy sets implies the introduction of another degree of freedom (non-memberships) into the set description. Such a generalization of fuzzy sets gives us an additional possibility to represent imperfect knowledge which leads to describing many real problems in a more adequate way. Applications of intuitionistic fuzzy sets to group decision making, negotiations, voting and other situations are presented in Szmidt and Kacprzyk [14], [16], [17], [20], [22], [24], [23], [25], [29], Szmidt and Kukier [32], [33]. (because of the different approaches presented in the works cited above, we are not able to discuss details here, and refer the interested reader directly to them).

## 2.1 Geometrical Representation

One of the possible geometrical representations of an intuitionistic fuzzy sets is given in Figure 1 (cf. Atanassov [3]). It is worth noticing that although we use a two-dimensional figure (which is more convenient to draw in our further considerations), we still adopt our approach (e.g., Szmidt and Kacprzyk [19], [28], [21], [30]), [31]) taking into account all three functions (membership, non-membership and hesitation margin values) describing an intuitionistic fuzzy set. Any element belonging to an intuitionistic fuzzy set may be represented inside an $MNO$ triangle. In other words, the $MNO$ triangle represents a surface where the coordinates of any element belonging to an A-IFS can be represented. Each point belonging to the $MNO$ triangle is described by the three coordinates: $(\mu, \nu, \pi)$. Points $M$ and $N$ represent crisp elements. Point $M(1,0,0)$ represents elements fully belonging to an A-IFS as $\mu = 1$, and may be seen as the representation of the ideal positive element. Point $N(0,1,0)$ represents elements fully not belonging to an A-IFS as $\nu = 1$, i.e. can be viewed as the ideal negative element. Point $O(0,0,1)$ represents elements about which we are not able to say if they belong or not belong to an A-IFS (intuitionistic fuzzy index $\pi = 1$). Such an interpretation is intuitively appealing and provides means for the representation of many aspects of imperfect information. Segment $MN$ (where $\pi = 0$) represents elements belonging to the classic fuzzy sets ($\mu + \nu = 1$). For example, point $A(0.5, 0.5, 0)$ (Figure 1), like any element from segment $MN$ represents an element of a fuzzy set. A line parallel to $MN$ describes the elements with the same values of the hesitation margin. In Figure 1 we can see point $B(0.4, 0.4, 0.2)$ representing an element with the hesitation margin equal 0.2, like $D(0.1, 0.7, 0.2)$, $E(0.5, 0.3, 0.2)$ and all elements on the line pointed out by any two from $B$, $E$, $D$. The closer a parallel line to $MN$ is to $O$, the higher the hesitation margin.

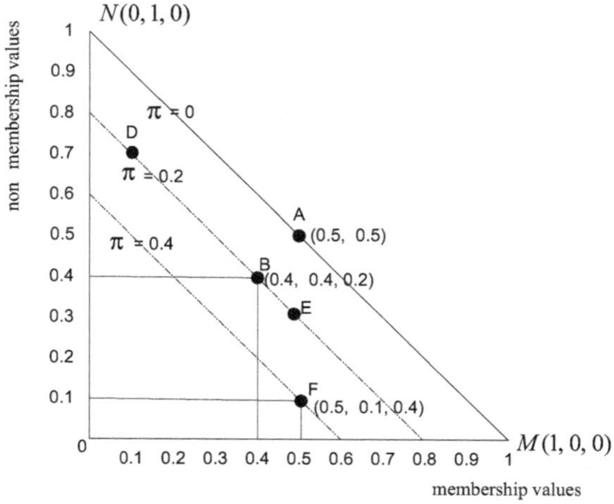


**Fig. 1.** Geometrical representation

**Remark:** We use the capital letters (e.g., $A$, $B$, $C$) for the geometrical representation of $x_i$'s (Figure 1) on the plane. The same abbreviations (capital letters) mean in this paper the sets but we always explain the current meaning of a symbol used.

### 2.2 Distances between the A-IFSs

In Szmidt and Kacprzyk [19], Szmidt and Baldwin [12,13], and especially in Szmidt and Kacprzyk [28] it is shown why while calculating distances between the A-IFSs we should take into account all three functions describing the A-IFSs. In [28] not only the reasons why we should take into account all three functions are given but also some possible serious problems that can occur while taking into account two functions only and that can imply some serious conceptual and numerical difficulties.

In our further considerations we will use the normalized Hamming distance between the A-IFSs $A, B$ in $X = \{x_1, \ldots, x_n\}$ (cf. Szmidt and Baldwin [12,13], Szmidt and Kacprzyk [19], [28]):

$$l_{IFS}(A, B) =$$
$$= \frac{1}{2n} \sum_{i=1}^{n} (|\mu_A(x_i) - \mu_B(x_i)| + |\nu_A(x_i) - \nu_B(x_i)| + |\pi_A(x_i) - \pi_B(x_i)|) \quad (6)$$

For (6) we have: $0 \leq l_{IFS}(A, B) \leq 1$. Clearly the normalized Hamming distance (6) satisfies the conditions of the metric.

## 3 Ranking the Alternatives

First, we will remind briefly some more relevant approaches known from the literature.

Chen and Tan [5] [1] proposed the concept of a score function for an intuitionistic fuzzy alternative $a = (\mu, \nu)$ meant as

$$S(a) = \mu - \nu, \tag{7}$$

and, clearly, $S(a) \in [-1, 1]$.

It follows immediately from (7) that the score function $S(a)$ alone is not enough for evaluating intuitionistic fuzzy alternatives as it produces the same result for such different intuitionistic fuzzy alternatives $a = (\mu, \nu)$ as, e.g.,: $(0.5, 0.4)$, $(0.4, 0.3)$, $(0.3, 0.2)$, $(0.1, 0)$ – for all of them $S(a) = 0.1$ which seems counterintuitive.

Then Hong and Choi [7] considered in addition to the score function as defined above, a so-called accuracy function $H$

$$H(a) = \mu + \nu, \tag{8}$$

where $H(a) \in [0, 1]$.

By making use of (7) and (8), Xu [34] proposed an algorithm ranking the intuitionistic fuzzy alternatives. We will present here its idea in the case of two alternatives $a_i$ and $a_j$ [34]:

- if $S(a_i) \leq S(a_j)$, then $a_i$ is smaller than $a_j$;
- if $S(a_i) = S(a_j)$, then:
  - if $H(a_i) = H(a_j)$, then $a_i$ and $a_j$ represent the same information (are equal);
  - if $H(a_i) \leq H(a_j)$, then $a_i$ is smaller than $a_j$.

However, the above ranking does not meet our expectation in many cases. Let us consider two intuitionistic fuzzy alternatives $a_1 = (0.5, 0.45)$ and $a_2 = (0.25, 0.05)$ for which we obtain $S(a_1) = 0.5 - 0.45 = 0.05$, $S(a_2) = 0.25 - 0.05 = 0.2$, suggesting that $a_1$ is smaller than $a_2$. But the information supplied by $a_1$ (i.e. $0.5 + 0.45 = 0.95$) is for sure greater than those supplied by $a_2$ (i.e. $0.25 + 0.05 = 0.3$). In other words, it is difficult to agree that $a_1$ is smaller than $a_2$. Later we will return to ranking the two intuitionistic fuzzy alternatives by the method we propose.

### 3.1 Ranking Alternatives via Distances from the Ideal Positive Alternative

In Section 2 we have mentioned some possible applications of the A-IFSs, among others, those related to voting. Now we will try to propose how to rank the voting alternatives expressed via intuitionistic fuzzy elements.

Let an element $x$ belonging to an A-IFS characterized via $(\mu, \nu, \pi)$ expresses a voting situation: $\mu$ means the proportion (from $[0, 1]$) of voters who vote for $x$, $\nu$ the proportion of those who vote against $x$, and $\pi$ of those who abstain. The simplest idea to compare different voting situations (ranking the alternatives) seems to use a distance measure from the ideal voting situation $M = (x, 1, 0, 0)$ (100% voting for, 0% vote against and 0% abstain) to the alternatives considered. We will call $M$ the ideal positive alternative.

[1] The score function in [5] is discussed for vague sets [6] but Bustince and Burillo [4] have proved that vague sets are equivalent to Atanassov's intuitionistic fuzzy sets.

Let:
$A = (x, 0.5, 0.5, 0)$ – 50% vote for, 50% against, and 0% abstain,
$B = (x, 0.4, 0.4, 0.2)$ – 40% vote for, 40% vote against and 20% abstain,
$C = (x, 0.3, 0.3, 0.4)$ – 30% vote for, 30% vote against and 40% abstain.

Certainly, the method of calculating distances between two A-IFSs $A$ and $B$ using the membership and non-membership values only (9) does not work properly (cf. Szmidt and Kacprzyk [19], [28], Szmidt and Baldwin [12], [13]) in this case, too:

$$l_2(A,B) = \frac{1}{2n}\sum_{i=1}^{n}(|\mu_A(x_i) - \mu_B(x_i)| + |\nu_A(x_i) - \nu_B(x_i)|) \tag{9}$$

The results from (9), i.e., the distances for the above voting alternatives represented by points $A, B, C$ (cf. Figure 2) from the ideal positive alternative represented by $M(1,0,0)$ are, respectively:

$$l_2(M,A) = 0.5(|1-0.5| + |0-0.5|) = 0.5 \tag{10}$$
$$l_2(M,B) = 0.5(|1-0.4| + |0-0.4|) = 0.5 \tag{11}$$
$$l_2(M,C) = 0.5(|1-0.3| + |0-0.3|) = 0.5 \tag{12}$$

The results seems to be counterintuitive as (9) suggests that all the alternatives (represented by) $A, B, C$ seem to be "the same". On the other hand, the normalized Hamming distance (6) taking into account besides the membership and non-membership the hesitation margin too, gives:

$$l_{IFS}(M,A) = 0.5(|1-0.5| + |0-0.5| + |0-0|) = 0.5 \tag{13}$$
$$l_{IFS}(M,B) = 0.5(|1-0.4| + |0-0.4| + |0-0.2|) = 0.6 \tag{14}$$
$$l_{IFS}(M,C) = 0.5(|1-0.3| + |0-0.3| + |0-0.4|) = 0.7 \tag{15}$$

The results (13)–(15) seem to reflect our intuition: alternative $A$ seems to be the best in the sense that the distance $l_{IFS}(M,A)$ is the smallest (we know for sure that 50% vote for, 50% vote against). The situation is given in Figure 2. The alternative represented $A$ is just a fuzzy alternative ($A$ lies on $MN$ where the values of the hesitation margin are equal 0). On the other hand, alternatives $B$ and $C$ are "less sure" (with the hesitation margin equal 0.2, and 0.4, respectively).

However, a weak point in the ranking of alternatives by calculating the distances from the ideal positive alternative represented by $M$ is that for a given value of the membership function, (6) gives just the same value (for example, if the membership value $\mu$ is equal 0.8, for any intuitionistic fuzzy element, i.e. such that its non-membership degree $\nu$ and hesitation margin $\pi$ fulfill $\nu + \pi = 0.2$, is equal 0.2). It is shown in Figure 3, a and b. To better see this, the distances (6) for any alternative from $M$ (Figure 3a) are presented for $\mu$ and $\nu$ for the whole range $[0,1]$ (instead for $\mu + \nu \leq 1$ only). For the same reason (to better see the effect), in Figure 3b) the contour plot of the distances (6) is given only for the range of $\mu$ and $\nu$ for which $\mu + \nu \leq 1$).

The conclusion is that the distances from the ideal positive alternative alone do not make it possible to rank the alternatives in the intended way.

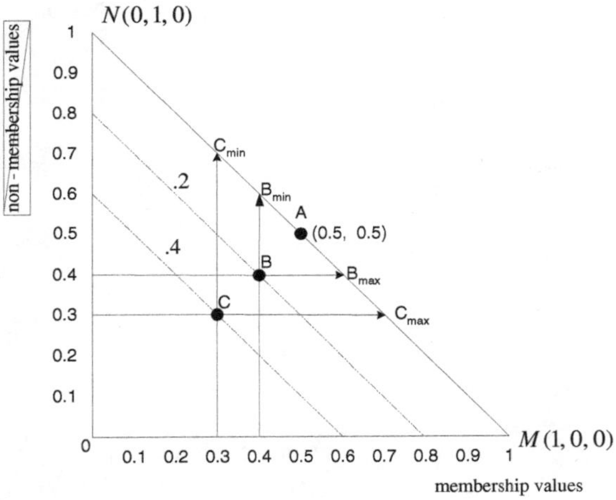


**Fig. 2.** Geometrical representation of IFSs

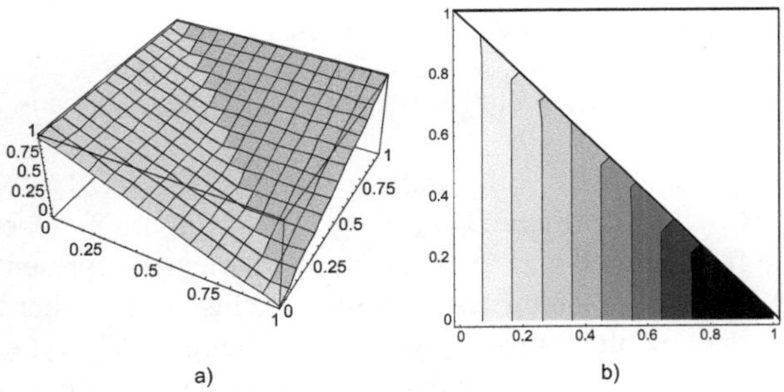


**Fig. 3.** a) Distances (6) of any IFS element from ideal alternative $M$; b) contour plot

### 3.2 A New Method for Ranking Alternatives

Let us analyze the sense of a voting alternative (expressed via an intuitionistic fuzzy element) using the operators of (cf. Atanassov [3]): *necessity* ( $\Box$ ), *possibility* ($\Diamond$), $D_\alpha(A)$ and $F_{\alpha,\beta}(A)$ given as:

- The *necessity* operator ( $\Box$ )

$$\Box A = \{\langle x, \mu_A(x), 1 - \mu_A(x)\rangle | x \in X\} \quad (16)$$

- The *possibility* operator ($\Diamond$)

$$\Diamond A = \{\langle x, 1 - \nu_A(x), \nu_A(x)\rangle | x \in X\} \quad (17)$$

- Operator $D_\alpha(A)$ (where $\alpha \in [0, 1]$)

$$D_\alpha(A) = \{\langle x,\ \mu_A(x) + \alpha\pi_A(x),\ \nu_A(x)(1-\alpha)\pi_A(x)\rangle \, | x \in X\} \quad (18)$$

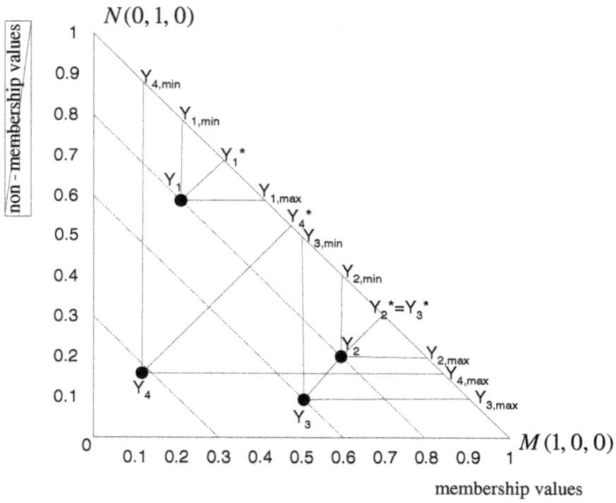


**Fig. 4.** Ranking alternatives $Y_i$

- Operator $F_{\alpha,\beta}(A)$ (where $\alpha, \beta \in [0,1]$; $\alpha + \beta \leq 1$)

$$F_{\alpha,\beta}(A) = \{\langle x,\ \mu_A(x) + \alpha\pi_A(x), \nu_A(x)\beta\pi_A(x)\rangle\, |x \in X\} \tag{19}$$

For example, for alternative $B(0.4, 0.4, 0.2)$ we obtain $\Box B = B_{min}$, where $B_{min} = (0.4, 0.6)$, and $\Diamond B = B_{max}$, where $B_{max} = (0.6, 0.4)$ (Figure 2). Operator $F_{\alpha,\beta}(A)$ makes it possible for alternative $B$ to become any alternative represented in triangle $BB_{max}B_{min}$. A similar reasoning leads to the conclusion that alternative $C$ (Figure 2) might become any alternative represented in triangle $CC_{max}C_{min}$, and alternative $O(0, 0, 1)$ (because of the hesitation margin equal 1) may become any alternative (the whole area of the triangle $MNO$).

Having the above considerations in mind we could say that the smaller the area of the triangle $Y_iY_{i,min}Y_{i,max}$ (Figure 4) the better alternative $Y_i$ from a set $Y$ of the alternatives considered. Alternatives having their representations on segment $MN$ are the best in the sense that:

- the hesitation margin is equal 0 here, which means that the alternatives are fully reliable in the sense of the information represented, and
- the alternatives are ordered – the closer an alternative to ideal positive alternative $M(1, 0, 0)$, the better it is (it is an obvious fact as fuzzy alternatives are univocally ordered).

The above reasoning suggests that a promising way of ranking the intuitionistic fuzzy alternatives $Y_i$ with the same values of $\pi_i$ is converting them into the fuzzy alternatives (which may be easily ranked). For alternatives $Y_i$ with different values of $\pi_i$ the simplest way to rank the alternatives seems to be to use information carried by triangles $Y_iY_{i,min}Y_{i,max}$.

$Y_i^*$ indicates the amount of information connected with $Y_i$ (the amount of information is indicated by "the position" of triangle $Y_iY_{i,min}Y_{i,max}$ inside triangle

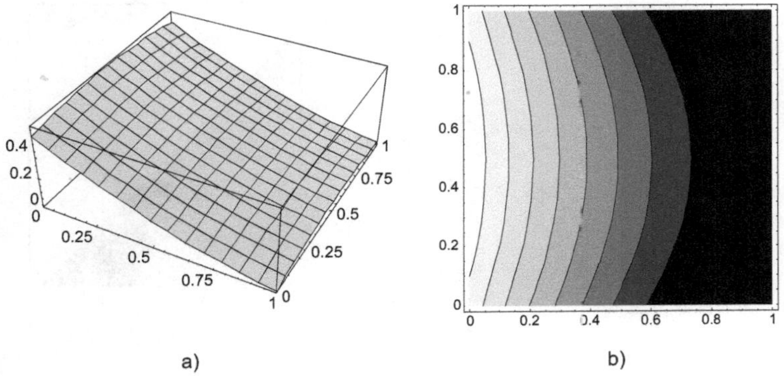


**Fig. 5.** a) $R(Y_i)$ as a function of a distance $Y_i^*$ from $M$ and a hesitation margin; b) contour plot

$MNO$ – expressed by the projection on segment $MN$). The value of the hesitation margin $\pi_{Y_i}$ indicates how sure (reliable) is the information represented by $Y_i^*$.

$Y_i^*$ are the orthogonal projections of $Y_i$ on $MN$. Szmidt and Kacprzyk [18] considered such an orthogonal projection of the intuitionistic fuzzy elements belonging to an intuitionistic fuzzy set $A$. This orthogonal projection may be obtained via operator $D_\alpha(A)$ (18) with parameter $\alpha$ equal 0.5.

It is worth noticing that all the elements from segment $OA$ (Figure 2) are transformed by $D_{0.5}(A)$ (18) into $A(0.5, 0.5)$ which reflects a lack of differences between the membership and non-membership, no matter which the value of the hesitation margin is.

In this context, a reasonable measure $R$ that can be used for ranking the alternatives (represented by) $Y_i$ seems to be

$$R(Y_i) = 0.5(1 + \pi_{Y_i})l_{IFS}(M, Y_i^*) \tag{20}$$

where $l_{IFS}(M, Y_i^*)$ is the distance (6) from the ideal positive alternative $M(1,0,0)$, $Y_i^*$ is the orthogonal projection of $Y_i$ on MN. Constant 0.5 was introduced in (20) to ensure that $0 < R(Y_i) \leq 1$. The values of function $R$ for any intuitionistic fuzzy element are presented in Figure 5a, and the counterpart contour plot – in Figure 5b. Unfortunately, the results obtained (20) do not rank the alternatives in the intended way. (The maximum value of (20) is not obtained for the alternative $(0, 0, 1)$ but for $(0, 1/2, 1/2)$.)

A better measure $R$ that can be used for ranking the alternatives (represented by) $Y_i$ seems to be

$$R(Y_i) = 0.5(1 + \pi_{Y_i})l_{IFS}(M, Y_i) \tag{21}$$

where $l_{IFS}(M, Y_i)$ is the distance (6) $Y_i$ from ideal positive alternative $M(1,0,0)$.

Equation (21) tells us about the "quality" of an alternative – the lower the value of $R(Y_i)$, (21), the better the alternative in the sense of the amount of positive information included, and reliability of information.

The best is alternative $M(1,0,0)$ for which $R(M) = 0$. For alternative $N(0,1,0)$ we obtain $R(N) = 0.5$ (alternative $N$ is fully reliable as the hesitation margin is equal 0, but the distance $l_{IFS}(M, N) = 1$). Alternative $A$ (Figures 1, 2) gives $R(A) = 0.25$.

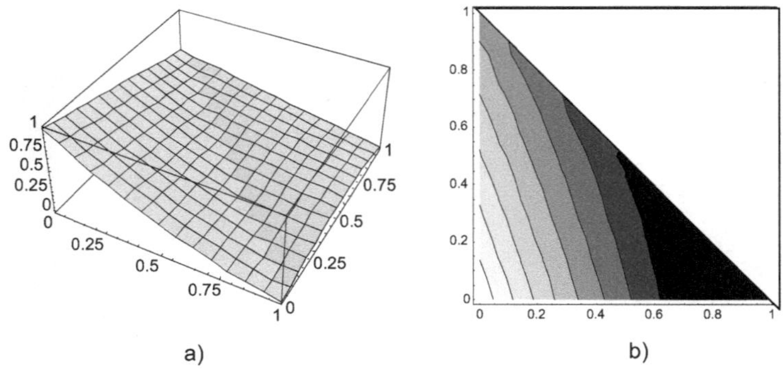


**Fig. 6.** a) $R(Y_i)$ as a function of a distance $Y_i$ from $M$ and a hesitation margin; b) contour plot

In general, on $MN$, the values of $R$ decrease from 0.5 (for alternative $N$) to 0 (for the best alternative $M$). The maximal value of $R$, i.e. 1, we obtain for $O(0,0,1)$ for which both the distance from $M$ and hesitation margin are equal 1 (alternative $O$ "indicates" the whole triangle $MNO$). All other alternatives $Y_i$ "indicate" smaller triangles $Y_iY_{i,min}Y_{i,max}$ (Figure 4), so their counterpart values of $R$ are smaller (better in the sense of the amount of the reliable information).

The values of function $R$ (21) for any intuitionistic fuzzy element are presented in Figure 6a, and the counterpart contour plot – in Figure 6b. Considering the numbers obtained via $R$ (21), we may notice that the value 0.25 obtained for the alternative (0.5, 0.5, 0) constitutes the "border" of the "interesting" alternatives – in the sense of the amount of the positive knowledge.

Let us return to the ranking of two alternatives (which were ranked counter-intuitively by the algorithm presented in [34] as shown in the beginning of Section 3), i.e., $Y_1 = (0.5, 0.45, 0.05)$ and $Y_2 = (0.25, 0.05, 0.7)$ (we stress here that we take into account all three values: the degrees of membership, non-membership and hesitation margin). From (21) we obtain: $R(Y_1) = 0.26$, $R(Y_2) = 0.64$ which means that $Y_1$ is better than $Y_2$ (previously, from the algorithm [34] $Y_2$ was better/bigger than $Y_1$). Obviously, $Y_1$ is not a "good" option as $R(Y_1)$ is bigger than 0.25 which follows from the fact that the non-membership value is quite big (equal 0.45). It might mean that we would not accept option $Y_1$. But option $Y_2$ seems even less interesting – with the smaller membership value (equal 0.25 instead of 0.5 for $Y_1$ ), and with the bigger hesitation margin (equal 0.7 instead of 0.05 for $Y_1$).

*Example 1.* Let us consider the ranking of six medical treatments, $C1 - C6$, affecting a patient in the following way:

- $C1 : (0.6, 0.2, 0.2)$ – influences in a positive way 60% of symptoms, in a negative way – 20% of symptoms, and its impact is unknown (was not confirmed) in a case of 20% of symptoms;
- $C2 : (0.7, 0.3, 0)$ – influences in a positive way 70% symptoms, in a negative way – 30% of symptoms, and its impact is unknown (was not confirmed) in case of 0% of symptoms;

**Table 1.** Ranking alternatives by $R$ (21) – results for the data from Example1

| No. | $C_i : (\mu_i, \nu_i, \pi_i)$ | $R_E(C_i)$ |
|---|---|---|
| 1 | $C1 : (0.6, 0.2, 0.2)$ | 0.240 |
| 2 | $C2 : (0.7, 0.3, 0)$ | 0.150 |
| 3 | $C3 : (0.7, 0.15, 0.15)$ | 0.173 |
| 4 | $C4 : (0.775, 0.225, 0)$ | 0.113 |
| 5 | $C5 : (0.8, 0.1, 0.1)$ | 0.110 |
| 6 | $C6 : (0.8, 0.2, 0)$ | 0.100 |

- $C3 : (0.7, 0.15, 0.15)$ – influences in a positive way 70% of symptoms, in a negative way – 15% of symptoms, and its impact is unknown (was not confirmed) in case of 15% of symptoms;
- $C4 : (0.775, 0.225, 0)$ – influences in a positive way 77.5% of symptoms, in a negative way – 22.5% of symptoms, and its impact is unknown (was not confirmed) in case of 0% of symptoms;
- $C5 : (0.8, 0.1, 0.1)$ – influences in a positive way 80% of symptoms, in a negative way – 10% of symptoms, and its impact is unknown (was not confirmed) in case of 10% of symptoms;
- $C6 : (0.8, 0.2, 0)$ – influences in a positive way 80% of symptoms, in a negative way – 20% of symptoms, and its impact is unknown (was not confirmed) in case of 0% of symptoms;

The ranking of $C1, \ldots, C6$ from (21) is given in Table 1 – from the worst one, $C1$ to the best one, $C6$.

In general, the ranking function $R$ (21) is constructed by strongly taking into account the lack of knowledge. Let us consider the pair: $C1$ and $C2$. In the case of $C1$ the lack of knowledge is 0.2, so that theoretically, we might expect "on the average" that the hesitation margin representing the lack of knowledge will be divided equally between the membership function and non-membership function giving as a result the case $C2$. But if we wish to avoid the most disadvantageous cases, we will rank $C2$ higher so as to avoid the possibility which might by implied by $C1$, namely: $(0.6, 0.4, 0)$ (while all the hesitation margin is added to the non-membership function). The best result which could happen (if the hesitation margin is added to the membership function of $C1$), namely $(0.8, 02, 0)$, (i.e. case $C6$ ranked as the best one – $R(C6) = 0.1$) does not influence the ranking of $C1(21)$.

Just the same situation can be observed for the pairs: $C3$ and $C4$, and next for $C5$ and $C6$. The existence of the non-zero hesitation margin influences negatively the ranking.

The obtained results seem to meet our expectations pretty well.

## 4 Conclusions

We have proposed a new method of ranking intuitionistic fuzzy alternatives. The method takes into account the amount of the information (both positive and negative) associated with an alternative (measured by a distance to the positive ideal alternative), and how reliable the information is (which is measured by the alternative's hesitation margin).

## References


1. Atanassov, K.: Intuitionistic Fuzzy Sets. VII ITKR Session. Sofia (Centr. Sci.-Techn. Libr. of Bulg. Acad. of Sci., 1697/84) (1983) (in Bulgarian)
2. Atanassov, K.: Intuitionistic Fuzzy Sets. Fuzzy Sets and Systems 20, 87–96 (1986)
3. Atanassov, K.: Intuitionistic Fuzzy Sets: Theory and Applications. Springer, Heidelberg (1999)
4. Bustince, H., Burillo, P.: Vague sets are intuitionistic fuzzy sets. Fuzzy Sets and Systems 67, 403–405 (1996)
5. Chen, S.M., Tan, J.M.: Handling multi-criteria fuzzy decision-making problems based on vague-set theory. Fuzzy Sets and Systems 67(2), 163–172 (1994)
6. Gau, W.L., Buehrer, D.J.: Vague sets. IEEE Trans. Systems Man Cybernet 23, 610–614 (1993)
7. Hong, D.H., Choi, C.H.: Multicriteria fuzzy decision making problems based on vague set theory. Fuzzy Sets and Systems 114, 103–113 (2000)
8. Li, F., Lu, A., Cai, L.: Methods of multi-criteria fuzzy decision making base on vague sets. J. of Huazhong Univ. of Science and Technology 29(7), 1–3 (2001)
9. Li, F., Rao, Y.: Weighted methods of multi-criteria fuzzy decision making based on vague sets. Computer Science 28(7), 60–65 (2001)
10. Hua-Wen, L., Guo-Jun, W.: Multi-criteria decision making methods based on intuitionistic fuzzy sets. European Journal of Operational Research 179, 220–233 (2007)
11. Rakus-Andersson, E.: Fuzzy and rough techniques in medical diagnosis and medication. Springer, Heidelberg (2007)
12. Szmidt, E., Baldwin, J.: New Similarity Measure for Intuitionistic Fuzzy Set Theory and Mass Assignment Theory. Notes on IFSs 9(3), 60–76 (2003)
13. Szmidt, E., Baldwin, J.: Entropy for Intuitionistic Fuzzy Set Theory and Mass Assignment Theory. Notes on IFSs 10(3), 15–28 (2004)
14. Szmidt, E., Kacprzyk, J.: Remarks on some applications of intuitionistic fuzzy sets in decision making. Notes on IFS 2(3), 22–31 (1996c)
15. Szmidt, E., Kacprzyk, J.: On measuring distances between intuitionistic fuzzy sets. Notes on IFS 3(4), 1–13 (1997)
16. Szmidt, E., Kacprzyk, J.: Group Decision Making under Intuitionistic Fuzzy Preference Relations. In: IPMU 1998, pp. 172–178 (1998a)
17. Szmidt, E., Kacprzyk, J.: Applications of Intuitionistic Fuzzy Sets in Decision Making. In: EUSFLAT 1999, pp. 150–158 (1998b)
18. Szmidt, E., Kacprzyk, J.: A Fuzzy Set Corresponding to an Intuitionistic Fuzzy Set. Int. Journal of Uncertainty Fuzziness and Knowledge Based Systems 6(5), 427–435 (1998c)
19. Szmidt, E., Kacprzyk, J.: Distances between intuitionistic fuzzy sets. Fuzzy Sets and Systems 114(3), 505–518 (2000)
20. Szmidt, E., Kacprzyk, J.: On Measures on Consensus Under Intuitionistic Fuzzy Relations. In: IPMU 2000, pp. 1454–1461 (2000)
21. Szmidt, E., Kacprzyk, J.: Entropy for intuitionistic fuzzy sets. Fuzzy Sets and Systems 118(3), 467–477 (2001)
22. Szmidt, E., Kacprzyk, J.: Analysis of Consensus under Intuitionistic Fuzzy Preferences. In: Proc. Int. Conf. in Fuzzy Logic and Technology, Leicester, UK, pp. 79–82 (2001)
23. Szmidt, E., Kacprzyk, J.: An Intuitionistic Fuzzy Set Based Approach to Intelligent Data Analysis (an application to medical diagnosis). In: Abraham, A., Jain, L., Kacprzyk, J. (eds.) Recent Advances in Intelligent Paradigms and Applications, pp. 57–70. Springer, Heidelberg (2002)

24. Szmidt, E., Kacprzyk, J.: Analysis of Agreement in a Group of Experts via Distances Between Intuitionistic Fuzzy Preferences. In: IPMU 2002, Annecy, France, pp. 1859–1865 (2002)
25. Szmidt, E., Kacprzyk, J.: An Intuitionistic Fuzzy Set Based Approach to Intelligent Data Analysis (an application to medical diagnosis). In: Abraham, A., Jain, L., Kacprzyk, J. (eds.) Recent Advances in Intelligent Paradigms and Applications, pp. 57–70. Springer, Heidelberg (2002b)
26. Szmidt, E., Kacprzyk, J.: Evaluation of Agreement in a Group of Experts via Distances Between Intuitionistic Fuzzy Sets. In: Proc. IS 2002 – Int. IEEE Symposium: Intelligent Systems, Varna, Bulgaria, pp. 166–170 (2002c)
27. Szmidt, E., Kacprzyk, J.: A New Concept of a Similarity Measure for Intuitionistic Fuzzy Sets and its Use in Group Decision Making. In: Torra, V., Narukawa, Y., Miyamoto, S. (eds.) MDAI 2005. LNCS (LNAI), vol. 3558, pp. 272–282. Springer, Heidelberg (2005)
28. Szmidt, E., Kacprzyk, J.: Distances Between Intuitionistic Fuzzy Sets: Straightforward Approaches may not work. In: 3rd Int. IEEE Conf. Intelligent Systems IS 2006, London, pp. 716–721 (2006)
29. Szmidt, E., Kacprzyk, J.: An Application of Intuitionistic Fuzzy Set Similarity Measures to a Multi-criteria Decision Making Problem. In: Rutkowski, L., Tadeusiewicz, R., Zadeh, L.A., Żurada, J.M. (eds.) ICAISC 2006. LNCS (LNAI), vol. 4029, pp. 314–323. Springer, Heidelberg (2006)
30. Szmidt, E., Kacprzyk, J.: Some problems with entropy measures for the Atannassov intuitionistic fuzzy sets. In: Masulli, F., Mitra, S., Pasi, G. (eds.) WILF 2007. LNCS (LNAI), vol. 4578, pp. 291–297. Springer, Heidelberg (2007)
31. Szmidt, E., Kacprzyk, J.: A New Similarity Measure for Intuitionistic Fuzzy Sets: Straightforward Approaches may not work. In: 2007 IEEE Conference on Fuzzy Systems, pp. 481–486 (2007a)
32. Szmidt, E., Kukier, M.: Classification of Imbalanced and Overlapping Classes using Intuitionistic Fuzzy Sets. In: 3rd International IEEE Conference on Intelligent Systems IS 2006, pp. 722–727 (2006)
33. Szmidt, E., Kukier, M.: A New Approach to Classification of Imbalanced Classes via Atanassov's Intuitionistic Fuzzy Sets. In: Wang, H.-F. (ed.) Intelligent Data Analysis: Developing New Methodologies Through Pattern Discovery and Recovery, pp. 85–101. IGI Global (2008)
34. Xu, Z.: Intuitionistic preference relations and their application in group decision making. Information Sciences 177, 2363–2379 (2007)
35. Zadeh, L.A.: Fuzzy sets. Information and Control 8, 338–353 (1965)

# Fuzzy Rule Base Model Identification by Bacterial Memetic Algorithms

János Botzheim[1,2], László Gál[3,4], and László T. Kóczy[2,3]

[1] Department of Automation, Széchenyi István University, Győr, Hungary
[2] Department of Telecommunication and Media Informatics, Budapest University of Technology and Economics, Budapest, Hungary
[3] Institute of Information Technology and Electrical Engineering, Széchenyi István University, Győr, Hungary
[4] Department of Technology, Informatics and Economy, University of West Hungary, Szombathely, Hungary
gallaci@ttmk.nyme.hu,
{botzheim,koczy}@{sze,tmit.bme}.hu

**Abstract.** Fuzzy systems have been successfully used in the area of controllers for a long time. The Mamdani method is one of the most popular inference systems for practical applications. The main problem of Mamdani-type inference system and other fuzzy logic based controllers is how to gain the fuzzy rules the inference system based on. Several approaches have been proposed for automatic rule base identification. The bacterial type evolutionary algorithms have been successfully applied for solving this task. These algorithms are based on the Pseudo-Bacterial Genetic Algorithm and are supplied with operations and methods (e.g. the Levenberg-Marquardt method) to complete their task more efficiently. The goal is to create more accurate fuzzy rule bases from input-output data sets as quickly as possible. In this work, we summarize the bacterial type evolutionary algorithms used for fuzzy rule base identification.

**Keywords:** Fuzzy systems, fuzzy rule base identification, bacterial type evolutionary algorithms, Levenberg-Marquardt method, memetic algorithms, Bacterial Memetic Algorithm, Modified Bacterial Memetic Algorithm.

## 1 Introduction

The evolution of the *evolutionary algorithms* started with the presentation of the *Genetic Algorithm* (GA) developed by J. H. Holland in 1975 [8]. These new types of algorithms were able to solve problems where analytic methods are not available or applying them could not present the result within a reasonable time. Although the *Genetic Algorithm* was not able to gain the optimal solution in the practice and it also has other drawbacks, it can be used to get *quasi-optimal* solutions in an *acceptable amount of time*.

In GA the parameters of a modeled system are encoded in a unit, named *chromosome*. Rather than trying to improve one parameter set encoded in one *chromosome* it

E. Rakus-Andersson et al. (Eds.): Recent Advances in Decision Making, SCI 222, pp. 21–43.
springerlink.com

is beneficial to have many of them and try to improve them in the way we have seen in the nature – the best individuals are staying alive and are able to reproduce themselves and inherit the information needed to survive encoded in their *chromosomes*. The operations in the original GA used in computing were named and operated the same way as in the nature (*mutation, crossover, selection*).

The way GA works it is mainly a global search in a certain search space to find better solutions like existing ones using the operation "*mutation*" that can change the parameters mostly randomly.

Theoretically the possibility of finding the optimal solution exists. Although in the practice we get acceptable solutions and the quality of the solutions is increasing as the processing time goes on, the convergence to the optimum is slow and decreases as the quality of the solution achieved is increasing.

One of the serious problems of *fuzzy rule base* modeling is how to find the optimal *fuzzy rule base* (FRB) when no human expert is available to gain the rules. In this case we need a method to identify the rule base automatically. Evolutionary algorithms are good candidates for this task because the parameters of a *fuzzy rule base* can be easily encoded in the *chromosome* of an *individual*, and the quality of the model achieved can be easily calculated if training data sets are present for a certain model. All we need are input-output data sets to develop models.

*Fuzzy systems* have been successfully used in the area of controllers for a long time. The first appearance of these controllers was in 1974 by Mamdani and Assilian [17]. The main problem in the usage of *Mamdani-type inference system* and other fuzzy logic based controllers is how to gain the *fuzzy rule base* what the inference system based on.

In 1997 N. E. Nawa, T. Hashiyama, T. Furuhashi and Y. Uchikawa proposed a new kind of evolutionary algorithm called *Pseudo-Bacterial Genetic Algorithm* (PBGA) [20] and applied successfully for *fuzzy rule base extraction* from input-output data sets. It introduces a new genetic operation called *bacterial mutation*. The power of this new operator can be utilized in environments where there are weak relationships between the parameters of the system. Fuzzy systems have this property. This new algorithm was simpler but more powerful used for *fuzzy rule base extraction* (faster convergence and better quality rules).

Furthermore, N. E. Nawa and T. Furuhashi proposed the *Bacterial Evolutionary Algorithm* (BEA) for fuzzy rule base extraction (1999) [21]. This new algorithm was based on PBGA supported by a new genetic operation called *gene transfer*. This new operation establishes relationships among the individuals of the population. It can also be used for decreasing or increasing the number of the rules in a fuzzy rule base.

Both the *Pseudo-Bacterial Genetic Algorithm* (PBGA) and the *Bacterial Evolutionary Algorithm* (BEA) are global search methods. The next remarkable step was the adaptation of the *Levenberg-Marquardt* method for *fuzzy rule base identification* (FRBI). In 2002 we applied the *Levenberg-Marquardt* method (LM) for FRBI successfully [3], as the derivatives for the Jacobian matrix can be computed for the general *trapezoidal fuzzy membership functions* (with COG defuzzification). The LM method is a local search algorithm, it provides a very fast convergence to the local optimum but is unable to avoid that.

In 2005 we proposed a more advanced approach the *Bacterial Memetic Algorithm* (BMA) [1]. This new algorithm combines the *Bacterial Evolutionary Algorithm* (as a global searcher part) and the *Levenberg-Marquard* method (as a local searcher). It provides significant improvements both in terms of the speed of convergence and in the quality of the model achieved in FRBI.

The BMA performed far better than its predecessors; however, there are some recent modifications to it for improve the performance. In 2008 we published some further improvements for the BMA. The improvements concern to the handling of the *knot order violation* that appears in the *Levenberg-Marquardt* method a part of the BMA used for fuzzy rule base extraction (*Improved BMA* – IBMA), and the modification of the operator execution order in the BMA for using the LM method more efficiently (*Bacterial Memetic Algorithm with the Modified Operator Execution Order* – BMAM). IBMA increases the speed of convergence rather for higher complexity fuzzy rule bases, while BMAM does the same rather for rule bases with lower complexity. Combining them the benefits of both methods can be utilized. The *Modified Bacterial Memetic Algorithm* (MBMA) combines the improvements of IBMA and BMAM.

This work summarizes the *bacterial type evolutionary algorithms* used for *fuzzy rule base identification* (FRBI).

## 2 Fuzzy systems

The *bacterial type evolutionary algorithms* have been first developed and applied for identifying *fuzzy rule bases* automatically.

In the course of the function of *fuzzy controllers* the input data is processed by the *inference system* supported by the so called *fuzzy rule base*. A *fuzzy rule base* consists of one or more *fuzzy rules*. One of such a *rule* holds the expected output for a certain input or inputs (in multidimensional case). Usually, in *fuzzy systems* the input and output data are not *crisp* but *fuzzy values* determined by a kind of *fuzzy membership function*. These *membership functions* can be rather different in the shape; however, in the practice the most commonly used ones are the *triangular shaped* and the *trapezoidal shaped fuzzy membership functions*.

Usually, in case of describing *triangular shaped membership functions* it is enough to specify two parameters (isosceles triangle): the position of the top (a) and the length of the base (b) of the triangle (Fig. 1).

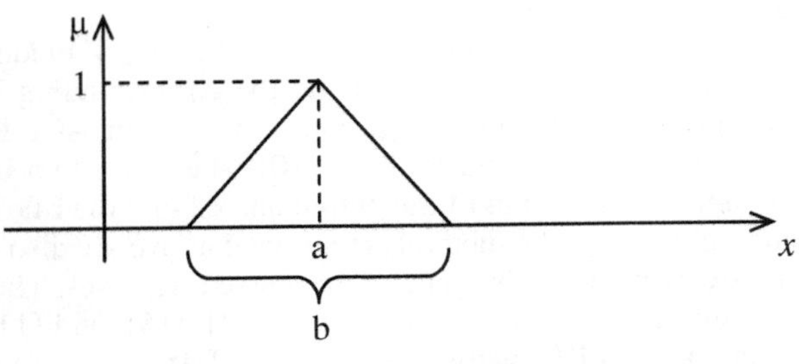


**Fig. 1.** Triangular shaped fuzzy membership function

μ denotes the *fuzzy membership function*, where $\mu(x_o)$ measures the applicability of the fuzzy rule for a certain input ( $x_o \in x$ ).

*Trapezoidal shaped fuzzy membership functions* offer more potential than the *triangular shaped* ones. These are widely used and are general enough from a practical point of view. Commonly, in case of describing *trapezoidal shaped fuzzy membership functions* four parameters are specified. These are the positions of the four *breakpoints* (a, b, c, d) of the trapezoid (Fig. 2).

In the case of using trapezoidal shaped fuzzy membership functions the four breakpoints that define the shape of each trapezoid must satisfy the following constraint:

$$a \le b \le c \le d \,. \tag{1}$$

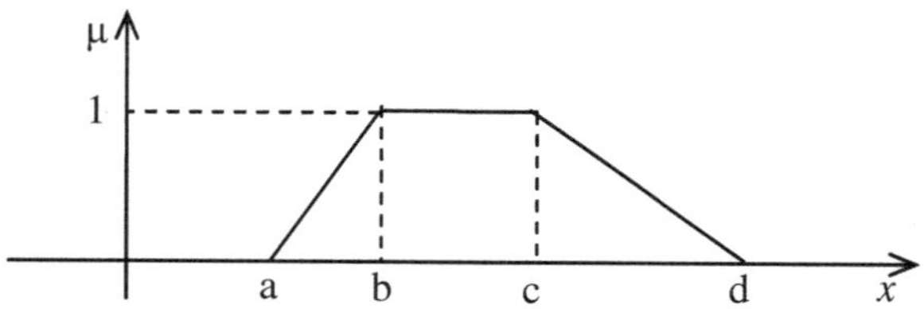


**Fig. 2.** Trapezoidal shaped fuzzy membership function

In case of *Mamdani-type inference systems* working with *trapezoidal shaped fuzzy membership functions*, one rule consists of $N_{Inputs}$ *antecedents* and one *consequent* (where $N_{Inputs}$ is the dimension of the input). Accordingly, since these are all trapezoids with four breakpoints, one rule can be defined by $(N_{Inputs}+1) \cdot 4$ parameters. As the *fuzzy rule base* contains $N_{Fuzzy\_rules}$ rules ($N_{Fuzzy\_rules}$ is the number of fuzzy rules of the rule base), the rule base can be defined by $N_{Fuzzy\_rules} \cdot (N_{Inputs} + 1) \cdot 4$ parameters. If a rule base is built up of 5 rules and the number of input variables is 6, then the total number of parameters ($N_{Parameters}$) needed to define the rule base is 140.

The task is to find the *fuzzy rule base* which one fits for the functioning of a given system best. In the case above it means determining and tuning of 140 parameters.

As we mentioned before one of the crucial problems of fuzzy rule base modeling is how to find the optimal or at least a quasi-optimal rule base for a certain system when no human expert is available to gain the rules. In this case we need a method for identifying the fuzzy rule base automatically.

One can find several approaches in the literature for fuzzy model identification (e.g. [22]). Some of them determine the rules and the corresponding linguistic terms based on fuzzy clustering (e.g. the method proposed in [13] or ACP in [9]). Another group of methods (e.g. RBE-DSS and RBE-SI [10]) start with two initial rules that describe the maximum and minimum of the output and extend the rule base iteratively in course of the tuning. Most of the methods mentioned above are also able to identify fuzzy models with low complexity by generating sparse rule bases. These systems use *fuzzy rule interpolation* (FRI) based reasoning (e.g. [14], [15], and [11]). An application oriented aspect of the FRI emerges in "FIVE" (Fuzzy Interpolation based on Vague Environment, originally introduced in [15], [16]), where for the sake of reasoning speed and direct real-time applicability the fuzziness of fuzzy partitions replaced by the concept of Vague Environment and hence the fuzzy interpolation to crisp one. Recently a freely available comprehensive FRI toolbox [12] and an FRI oriented web site (*fri.gamf.hu*) were appeared for aiding and guiding the future FRI applications.

Various evolutionary approaches have been proposed for fuzzy rule base extraction from input-output data such as the *Pseudo-Bacterial Genetic Algorithm* (PBGA) [20], the *Bacterial Evolutionary Algorithm* (BEA) [21], the *Bacterial Memetic Algorithm* (BMA) [1], the *Improved Bacterial Memetic Algorithm* (IBMA) [5], the *BMA with the Modified Operator Execution Order* (BMAM) [6] and the *Modified Bacterial Memetic Algorithm* (MBMA) [7]. All these have turned out to be helpful with the construction of such fuzzy rule base models; however, their respective optimality has been different in each case. In the next sections we discuss these *bacterial type evolutionary algorithms* used for *fuzzy rule base identification.*

## 3 Pseudo-Bacterial Genetic Algorithm (PBGA)

The original genetic algorithm (GA) was developed by Holland [8] and was based on the process of evolution of biological organisms. These processes can be easily applied in optimization problems where one individual corresponds to one solution of the problem.

Nawa, Hashiyama, Furuhashi and Uchikawa proposed a novel type of evolutionary algorithm called *Pseudo-Bacterial Genetic Algorithm* (PBGA) for fuzzy rule base extraction (1997) [20]. The *Pseudo-Bacterial Genetic Algorithm* is a special kind of *Genetic Algorithm* [8]. Its core contains a new genetic operation called *bacterial mutation*, which is inspired by the biological bacterial cell model, so this method mimics the microbial evolution phenomenon. Its basic idea is to improve the parts of *chromosomes* contained in each *bacterium.*

*Bacteria* can transfer *genes* to other *bacteria.* This mechanism is used in the *bacterial mutation.* For the *bacterial algorithm,* the first step is to determine how the problem can be encoded in a *bacterium* (*chromosome*). Our task is to find the optimal *fuzzy rule base* for a pattern set. Thus, the parameters of the *fuzzy rules* must be encoded in the *bacterium.* In the general case the parameters of the rules are the breakpoints of the trapezoids, thus, a bacterium will contain these breakpoints. For example, the encoding method of a fuzzy system with two inputs and one output can be seen in Fig. 3.

The next step is to optimize the parameters. Therefore a procedure is working on changing the parameters, testing the model obtained by this way and selecting the best models. In the course of testing the input-output data used for training are compared to the input and the output of the model (SSE, *MSE*, BIC). The smaller is the error, the better is the performance of the model. The *inference system* used for the model calculations can be any of the various types of *fuzzy inference systems.* Our investigations have been done mainly on Mamdani-type inference systems with trapezoidal shaped membership functions, as these are widely used and general enough from a mathematical point of view.

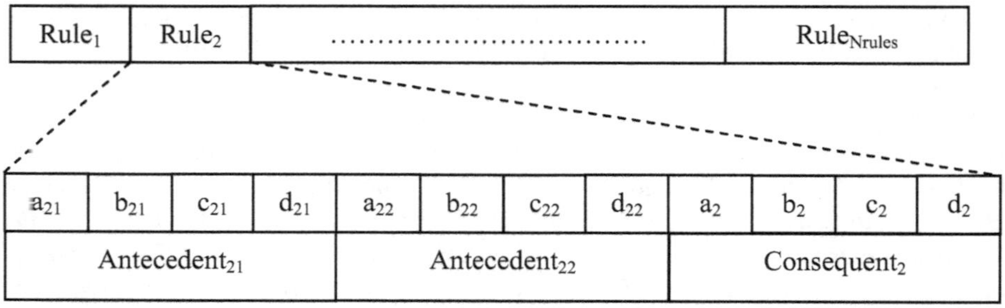


**Fig. 3.** Encoding of the fuzzy rules

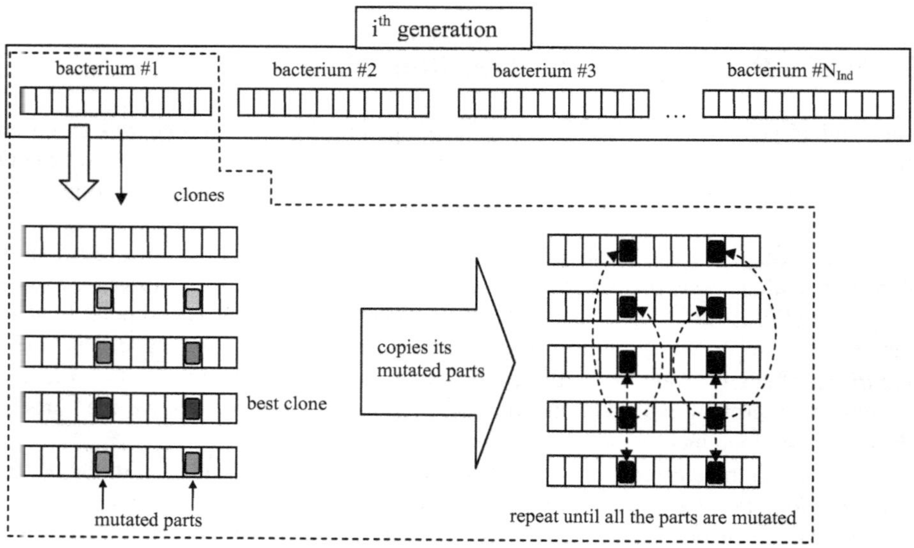


**Fig. 4.** Bacterial mutation (one individual)

The flowchart of the *Pseudo-Bacterial Genetic Algorithm* can be seen in Fig. 5, and its main steps are described below:

- Create the initial population: $N_{Ind}$ individuals are randomly created and evaluated. ($N_{Ind}$ is the number of individuals in the population.) Each individual contains $N_{Fuzzy\_rules}$ fuzzy rules encoded in the chromosome ($N_{Fuzzy\_rules}$ is the number of fuzzy rules of the desired model).
- Apply the *bacterial mutation* to each individual (Fig. 4)
  - Each individual is selected one by one.
  - $N_{Clones}$ copies of the selected individual are created ("clones").
  - Choose the same part or parts randomly from the clones and mutate it (except one single clone that remains unchanged during this mutation cycle).

- Select the best clone and transfer its mutated part or parts to the other clones.
- Repeat the part choosing-mutation-selection-transfer cycle until all the parts are mutated and tested exactly once.
- The best individual is remaining in the population, all other clones are deleted.
- This process is repeated until all the individuals have gone through the bacterial mutation.

- Apply conventional genetic operations (selection, reproduction and crossover).
- Repeat the procedure above from the *bacterial mutation* step until a certain termination criterion is satisfied (e.g. maximum number of generations).

Hints/recommendations:

- For the initial population the parameters are generated randomly in the interval of the range defined for the respective input.
- In the *bacterial mutation cycle* one selected *part* can be either a trapezoid or a breakpoint. We recommend the use of the breakpoint.
- It is recommended to choose the number of the parts selected for the bacterial mutation randomly each time because this method ensures the possibility of changing 1, 2 … $N_{Parameters}$ parameters at a time. In case of 1 or "low value" the *mutation* improves the parts of the FRB (which is the main idea of the *bacterial mutation*), while in case of higher numbers the search is done in the whole search space (especially when all parts are selected at a time), so the local optima can be more easily avoided.
- In the *bacterial mutation* sometimes it is not enough to generate a random number between the lower and upper boundary of the selected parameter's input range. The interval may be extended (i.e. by ±10 percent to the range or to the current value, or by a half of a transformed normal distribution on the lower and upper end of the interval.)
- In one iteration of full bacterial mutation (one generation) one part is selected for mutation exactly once.
- After the bacterial mutation, and before the model evaluation, the breakpoints of the trapezoid that contains the mutated part have to be ordered.

The algorithm works efficiently in environments where there are weak relationships between the parameters encoded in the chromosome. *Fuzzy rule bases* have this property, so PBGA has been successfully applied for obtaining quasi-optimal rules of fuzzy systems based on input-output training sets. PBGA performs well, converges fast towards the optimal rule base and it can be simply implemented.

## 4 Bacterial Evolutionary Algorithm (BEA)

*Bacterial Evolutionary Algorithm* (BEA) is based on the PBGA supported by a new genetic operation called *gene transfer operation* [21]. This new operation establishes relationships among the individuals of the population.

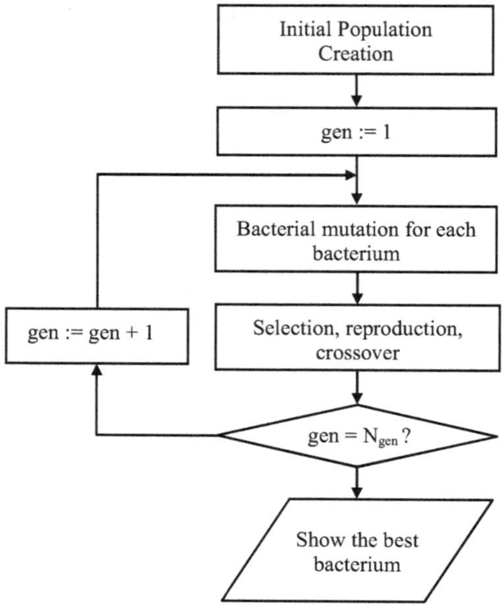


**Fig. 5.** Flowchart of the PBGA

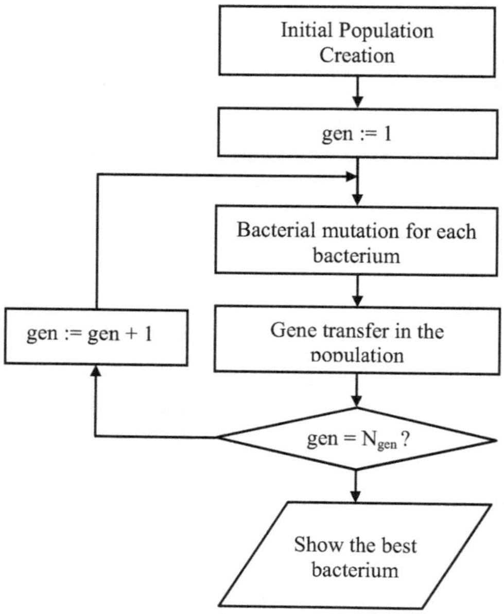


**Fig. 6.** Flowchart of the BEA

The main steps of the *gene transfer operation* are (Fig. 7):

- Sort the population according to the fitness values and divide it in two halves. The half that contains the better individuals is called "*superior half*" while the other half is the "*inferior half*".
- Choose one individual (the "*source chromosome*") from the *superior half* and another one (the "*destination chromosome*") from the *inferior half*.
- Transfer a part from the *source chromosome* to the *destination chromosome* (select the part randomly or by a predefined criterion).
- Repeat the steps above $N_{Inf}$ times ($N_{Inf}$ is the number of "*infections*" to occur in one generation.)

The *gene transfer operation* can be used in place of *selection, reproduction, crossover* in the algorithm described in Fig. 5. The flowchart of the *Bacterial Evolutionary* Algorithm can be seen in Fig. 6.

The *gene transfer* operation can also be used for decreasing or increasing the number of the rules in a fuzzy rule base (Fig. 7).

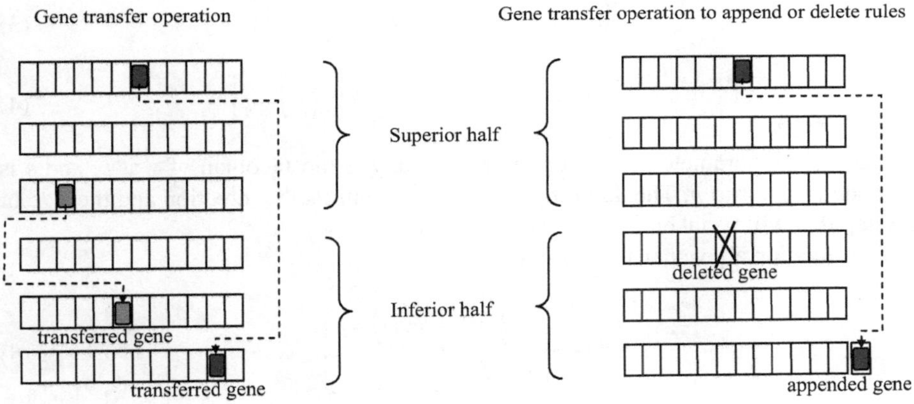


**Fig. 7.** Gene transfer operations

## 5 Bacterial Memetic Algorithm (BMA)

Memetic Algorithms combine evolutionary and local search methods (P. Moscato, 1989) [19]. The evolutionary part is able to find the global optimum region, but is not suitable to find the local minimum in practice. The gradient based part is able to reach the local optimum, but is very sensitive to the initial position in the search space and is unable to avoid the local optimum. Combining global and local search is expected to be beneficial.

*Bacterial Memetic Algorithm* (BMA) is a very recent approach (2005) [1]. It combines the *Bacterial Evolutionary Algorithm* and the *Levenberg-Marquardt* method. It can be used for fuzzy rule base identification because the derivatives for the Jacobian matrix can be computed for the general trapezoidal fuzzy membership functions

(*Mamdani-type inference system* with *COG defuzzification*) [1], [2], [4]. It provides significant improvements both in terms of the speed of convergence and in the quality of the model achieved in FRB identification.

### 5.1 Levenberg-Marquardt Method (LM)

The *Levenberg-Marquardt* (LM) method [18] is a gradient based iterative procedure. It is used for least squares curve fitting for a given set of empirical data ($\underline{x}_i$, $t_i$), minimizing the sum of squared error function (SSE)

$$E = \sum_{i=1}^{m} \left[ y_i(\underline{x}_i, \underline{p}) - t_i \right]^2 , \tag{2}$$

where $\underline{t}$ is the target vector, $\underline{y}$ is the output vector produced by the model, $\underline{x}$ is the input vector and $\underline{p}$ is the parameter vector to be optimized.

Its main equation is

$$\left( \underline{\underline{J}}^T \underline{\underline{J}} + \lambda \underline{\underline{I}} \right) \underline{s} = -\underline{\underline{J}}^T \left[ \underline{y}(\underline{p}) - \underline{t} \right], \tag{3}$$

$$\underline{s} = -\left( \underline{\underline{J}}^T \underline{\underline{J}} + \lambda \underline{\underline{I}} \right)^{-1} \underline{\underline{J}}^T \left[ \underline{y}(\underline{p}) - \underline{t} \right] \tag{4}$$

where $\underline{p}$ is the parameter vector to be optimized, $\underline{\underline{J}}$ is the Jacobian of $\underline{y}$ at $\underline{p}$, and $\underline{s}$ is the update vector to $\underline{p}$. The dumping parameter λ controls the direction and the size of the step that will be taken.

The equation above can be recast as

$$\underline{s} = -\begin{bmatrix} \underline{\underline{J}} \\ \sqrt{\lambda} \underline{\underline{I}} \end{bmatrix}^{+} \begin{bmatrix} \underline{y}(\underline{p}) - \underline{t} \\ \underline{0} \end{bmatrix} . \tag{5}$$

The operator $^+$ denotes the Moore-Penrose pseudoinverse. The uniform complexity of its accurate calculation is $O(p \cdot q \cdot \min(p \cdot q))$, where $p$ and $q$ are the number of rows and columns of the matrix, respectively. (If the rank-r of the matrix is low then the complexity can be reduced to $O(p \cdot q \cdot r)$, where $r \le \sqrt{\min(p,q)}$ [23].) Because the number of the parameters of a fuzzy rule base ($N_{parameters}$) is less than the number of patterns in the training data set ($N_{patterns}$) thus the uniform complexity of the LM algorithm used for FRBI is $O\left(N_{parameters}^2 \cdot N_{patterns}\right)$, which depends on the amount of the data linearly.

After solving the equation above in the k$^{th}$ iteration the update vector $\underline{s}$ is applied to optimize the parameter vector $\underline{p}$ in the following way:

$$\underline{p}[k] = \underline{p}[k-1] + \underline{s}[k] . \tag{6}$$

In case of FRB optimization the parameter vector contains the parameters of one FRB (or one *chromosome*).

The LM method alone can also be used for fuzzy rule extraction [2], but it generates only locally optimal rule base in the neighborhood of the initial rules.

### 5.1.1 The Jacobian Computation

The key in applying the LM method is how to get the LM update vector. To calculate the update vector the Jacobian matrix $\underline{\underline{J}}$ with respect to the parameters in the rules has to be computed. This will be done as shown below, in a pattern by pattern (*pt*) basis.

$$\underline{\underline{J}}[k] = \left[ \frac{\partial y(\underline{x}^{(pt)})[k]}{\partial \underline{p}[k]} \right], \tag{7}$$

where $k$ is the iteration variable. This can be written as follows:

$$\underline{\underline{J}} = \left[ \frac{\partial y\left(\underline{x}^{(pt)}\right)}{\partial a_{11}} \frac{\partial y\left(\underline{x}^{(pt)}\right)}{\partial b_{11}} \cdots \frac{\partial y\left(\underline{x}^{(pt)}\right)}{\partial a_{12}} \cdots \frac{\partial y\left(\underline{x}^{(pt)}\right)}{\partial d_{1}} \cdots \frac{\partial y\left(\underline{x}^{(pt)}\right)}{\partial d_{R}} \right], \tag{8}$$

where

$$\begin{aligned}
\frac{\partial y\left(\underline{x}^{(pt)}\right)}{\partial a_{ij}} &= \frac{\partial y}{\partial w_i}\frac{\partial w_i}{\partial \mu_{ij}}\frac{\partial \mu_{ij}}{\partial a_{ij}} \\
\frac{\partial y\left(\underline{x}^{(pt)}\right)}{\partial b_{ij}} &= \frac{\partial y}{\partial w_i}\frac{\partial w_i}{\partial \mu_{ij}}\frac{\partial \mu_{ij}}{\partial b_{ij}} \\
\frac{\partial y\left(\underline{x}^{(pt)}\right)}{\partial c_{ij}} &= \frac{\partial y}{\partial w_i}\frac{\partial w_i}{\partial \mu_{ij}}\frac{\partial \mu_{ij}}{\partial c_{ij}} \\
\frac{\partial y\left(\underline{x}^{(pt)}\right)}{\partial d_{ij}} &= \frac{\partial y}{\partial w_i}\frac{\partial w_i}{\partial \mu_{ij}}\frac{\partial \mu_{ij}}{\partial d_{ij}}
\end{aligned} \tag{9}$$

$w_i$ denotes the activation degree of the $i^{th}$ rule (the t-norm is the minimum):

$$w_i = \min_{j=1}^{n} \mu_{ij}\left(x_j\right) \tag{10}$$

where $n$ is the number of the input dimensions and $\mu_{ij}$ denotes the relative importance of the $j^{th}$ fuzzy variable in the $i^{th}$ rule.

$$\mu_{ij}(x_j) = \frac{x_j - a_{ij}}{b_{ij} - a_{ij}} N_{i,j,1}(x_j) + N_{i,j,2}(x_j) + \frac{d_{ij} - x_j}{d_{ij} - c_{ij}} N_{i,j,3}(x_j) \tag{11}$$

$$N_{i,j,1}(x_j)=\begin{cases}1, & if \quad x_j\in\lfloor a_{ij},b_{ij}\rfloor\\ 0, & otherwise\end{cases}$$
$$N_{i,j,2}(x_j)=\begin{cases}1, & if \quad x_j\in\lfloor b_{ij},c_{ij}\rfloor\\ 0, & otherwise\end{cases} \qquad (12)$$
$$N_{i,j,3}(x_j)=\begin{cases}1, & if \quad x_j\in\lfloor c_{ij},d_{ij}\rfloor\\ 0, & otherwise\end{cases}$$

The i[th] output is being cut in the height $w_i$, and with the *Center of Gravity* (COG) *defuzzification method* the output is calculated:

$$y(\underline{x})=\frac{\sum_{i=1}^{N_{FuzzRules}}\int_{y\in \operatorname{supp}\mu_i(y)} y\mu_i(y)dy}{\sum_{i=1}^{N_{FuzzRules}}\int_{y\in \operatorname{supp}\mu_i(y)} \mu_i(y)dy}. \qquad (13)$$

If this defuzzification method is used, the integrals can be easily computed:

$$y(x)=\frac{1}{3}\frac{\sum_{i=1}^{N_{FuzzRules}}(C_i+D_i+E_i)}{\sum_{i=1}^{N_{FuzzRules}}2w_i(d_i-a_i)+w_i^2(a_i-b_i+c_i-d_i)}$$
$$C_i=3w_i(d_i^2-a_i^2)(1-w_i) \qquad (14)$$
$$D_i=2w_i^2(c_id_i-a_ib_i)$$
$$E_i=w_i^3(c_i-d_i+a_i-b_i)(c_i-d_i-a_i+b_i)$$

It can be seen from (10) that $w_i$ depends on the membership functions, and each membership function depends only on four parameters (breakpoints). So, the derivatives of $w_i$ will be:

$$\frac{\partial w_i}{\partial \mu_{ij}}=\begin{cases}1, & if \quad \mu_{ij}=\min_{k=1}^{n}\mu_{ik}\\ 0, & otherwise\end{cases} \qquad (15)$$

The derivatives of the membership functions will be calculated as follows:

$$\frac{\partial \mu_{ij}}{\partial a_{ij}}=\frac{x_j^{(p)}-b_{ij}}{(b_{ij}-a_{ij})^2}N_{i,j,1}\left(x_j^{(p)}\right) \qquad (16)$$

$$\frac{\partial \mu_{ij}}{\partial b_{ij}} = \frac{a_{ij} - x_j^{(p)}}{\left(b_{ij} - a_{ij}\right)^2} N_{i,j,1}\left(x_j^{(p)}\right) \tag{17}$$

$$\frac{\partial \mu_{ij}}{\partial c_{ij}} = \frac{d_{ij} - x_j^{(p)}}{\left(d_{ij} - c_{ij}\right)^2} N_{i,j,3}\left(x_j^{(p)}\right) \tag{18}$$

$$\frac{\partial \mu_{ij}}{\partial d_{ij}} = \frac{x_j^{(p)} - c_{ij}}{\left(d_{ij} - c_{ij}\right)^2} N_{i,j,3}\left(x_j^{(p)}\right) \tag{19}$$

$\frac{\partial y}{\partial w_i}$ and the derivatives of the output membership functions' parameters have to be also computed. From (14) the following can be written:

$$\frac{\partial y}{\partial *_{i^*}} = \frac{1}{3} \frac{den \frac{\partial F_{i^*}}{\partial *_{i^*}} - num \frac{\partial G_{i^*}}{\partial *_{i^*}}}{(den)^2} \tag{20}$$

Where $*_i = w_i, a_i, b_i, c_i, d_i$ ; *den* is the denominator and *num* is the numerator of (14), resp. $F_{i^*}$ is the $i^*$ member of the sum in the numerator and $G_{i^*}$ is the $i^*$ member in the denominator. The derivatives will be given as follows:

$$\frac{\partial F_i}{\partial a_i} = -6w_i a_i + 6w_i^2 a_i - 3w_i^2 b_i - 2w_i^3 (a_i - b_i) \tag{21}$$

$$\frac{\partial G_i}{\partial a_i} = -2w_i + 2w_i^2 \tag{22}$$

$$\frac{\partial F_i}{\partial b_i} = -3w_i^2 a_i + 2w_i^3 (a_i - b_i) \tag{23}$$

$$\frac{\partial G_i}{\partial b_i} = -w_i^2 \tag{24}$$

$$\frac{\partial F_i}{\partial c_i} = 3w_i^2 d_i - 2w_i^3 (d_i - c_i) \tag{25}$$

$$\frac{\partial G_i}{\partial c_i} = w_i^2 \tag{26}$$

$$\frac{\partial F_i}{\partial d_i} = 6w_i d_i - 6w_i^2 d_i + 3w_i^2 c_i + 2w_i^3 (d_i - c_i) \tag{27}$$

$$\frac{\partial G_i}{\partial d_i} = 2w_i - w_i^2 \tag{28}$$

$$\frac{\partial F_i}{\partial w_i} = 3\left(d_i^2 - a_i^2\right)\left(1 - 2w_i\right) + 6w_i\left(c_i d_i - a_i b_i\right) + 3w_i^2\left[\left(c_i - d_i\right)^2 - \left(a_i - b_i\right)^2\right] \tag{29}$$

$$\frac{\partial G_i}{\partial w_i} = 2\left(d_i - a_i\right) + 2w_i\left(c_i + a_i - d_i - b_i\right) \tag{30}$$

The number of columns of $\underline{\underline{J}}$ is $4 \cdot (n+1) \cdot R$.

### 5.1.2 Knot Order Violation

In the BMA the LM procedure has to be modified. In case of trapezoidal shaped fuzzy membership function the parameter vector contains the four breakpoints (a, b, c, d or $K_1$, $K_2$, $K_3$, $K_4$) for each trapezoid. Applying the update vector calculated by the LM method some breakpoints of the trapezoids may be swapped. It happens not too often but it may cause serious problem as abnormal trapezoids may be obtained (Fig. 9). In case the order of the breakpoints of a trapezoid does not satisfy the $K_1 \le K_2 \le K_3 \le K_4$ constraint, then the membership function defined by the four breakpoints cannot be interpreted as a fuzzy membership function.

In the BMA in case of *knot order violation* (KOV) an *update vector reduction factor* is applied ($g$) [3]. This factor is a number between 0 and 1, it limits the magnitude of the update computed by LM for that pair of points which causes the damage of the knot order. It can be calculated as follows:

$$g = \frac{K_{i+1}[k-1] - K_i[k-1]}{2\left(s_i[k] - s_{i+1}[k]\right)}, \tag{31}$$

where $K_i[k-1]$ is the $i^{th}$ breakpoint of the trapezoid before the $k^{th}$ iteration (at the beginning of the current LM iteration), and $s_i[k]$ is the current LM update for the $i^{th}$ breakpoint. After calculating factor $g$ the adjusted position of the breakpoints can be computed as ($K'$):

$$\begin{aligned} K'_i[k] &:= K_i[k-1] + g \cdot s_i[k], \\ K'_{i+1}[k] &:= K_{i+1}[k-1] + g \cdot s_{i+1}[k]. \end{aligned} \tag{32}$$

This way the *knot order violation* can always be avoided.

## 5.2 The Bacterial Memetic Algorithm

Combining the *Levenberg-Marquardt* method modified for *fuzzy rule base identification* above with the *Bacterial Evolutionary Algorithm* provides definitely better results in *fuzzy rule base extraction.* The new algorithm is based on the operations of the PGBA (*bacterial mutation*), BEA (*gene transfer*) and the *Levenberg-Marquardt* method. It is much more successful in FRB identification than its predecessors.

The flowchart of the *Bacterial Memetic Algorithm* can be seen in Fig. 8, and its main steps are described below:

- Create the initial population.
- Apply the *bacterial mutation* to each individual.
- Apply the *Levenberg-Marquardt* method to each individual (e.g. 10 iterations per individual per generation).
- Apply the *gene transfer* operation $N_{Inf}$ times per generation.
- Repeat the procedure above from the *bacterial mutation* step until a certain termination criterion is satisfied.

The most significant improvement in the speed of convergence and the quality of the model achieved by the FRB identification process was the idea of combining the global and local search methods – the BMA.

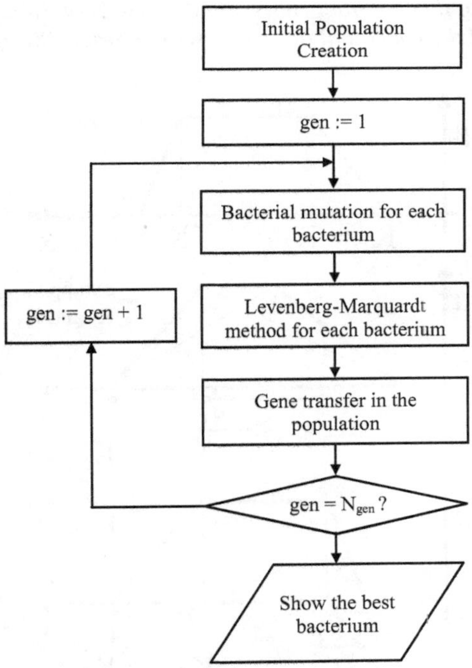


**Fig. 8.** Flowchart of the BMA

## 6 Improved Bacterial Memetic Algorithm (IBMA)

Although *Bacterial Memetic Algorithm* provides a very good speed of convergence towards the optimal rule base there are likely some points of the algorithm where the performance could be increased. One of these points concerns the *knot order violation handling*.

As mentioned before, the original BMA handles this problem by computing and applying the *update vector reduction factor*. The drawback of the above method is that in case of knot order's damage the full power of the LM method cannot be utilized because it limits the magnitude of the update (approx. to the half of the allowed value), and this method should be integrated into the LM procedure much deeper.

We proposed new elements (*Swap, Merge*) for KOV handling in LM used in BMA (2008) [5]. The algorithm containing a new KOV handling technique (*Swap)* rather than the *update vector reduction factor* is simpler and a slightly more powerful than the BMA. It is called *Improved Bacterial Memetic Algorithm* (IBMA).

### 6.1 Improvements in Knot Order Violation Handling

The two alternative methods for KOV handling are:

*a. Merge of the violating knots into a single knot. (Merge)*
*b. Swap of the knots that are in the wrong order. (Swap)*

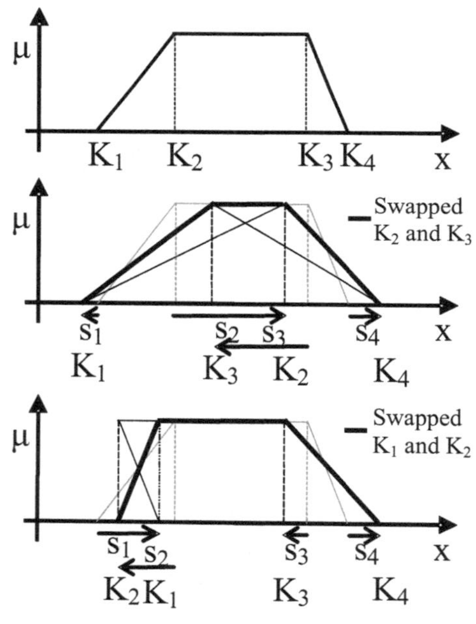


**Fig. 9.** KOVH method Swap

We found that both of these methods perform slightly better than the original one used in the BMA (especially the method *swap*), besides they are easier to implement and to integrate in the BMA.

The main point of the KOV handling method *swap* is in the case of the knot order violation the rate of the shift of both of the two breakpoints that have been computed by the LM method has to be applied as much as it can be done; however without the formation of trapezoids with vertical edges, and in such a manner that the algorithm can be applied *after* the update part of the LM algorithm. Corresponding to these, the recommended method is to *swap* the two violating knots, so the formation of abnormal trapezoids or trapezoids with vertical edges can always be avoided (Fig. 9).

## 7 BMA with the Modified Operator Execution Order (BMAM)

Another improvement of the BMA is the *Bacterial Memetic Algorithm with the Modified Operator Execution Order* (BMAM) (2008) [6]. This new approach exploits the Levenberg-Marquardt method more efficiently.

The BMA integrates its two components, the BEA and the LM method in the following way:

1. *Bacterial Mutation operation* for each individual,
2. *Levenberg-Marquardt* method for each individual,
3. *Gene Transfer operation* for a partial population.

This way the LM method is nested into the BEA, so that local search is done for every global search cycle.

Instead of applying the LM cycle *after* the *bacterial mutation* as a separate step, the modified algorithm executes several LM cycles *during* the *bacterial mutation* after *each mutational* step.

The *bacterial mutation* operation changes one or more breakpoints of the trapezoidal shaped fuzzy membership functions of a fuzzy rule base randomly, and then it tests whether the rule base obtained by this way performs better than the previous rule base or the rule bases that have been changed concurrently this way in the other so called clones. The mutation test cycle is repeated until all the parameters of the rule base have gone through the bacterial mutation.

In the mutational cycle it is possible to gain a rule base that has an instantaneous fitness value that is worse than the one in the previous or the concurrent rule bases. However, it is potentially better than those, because it is located in such a region of the search space which has a better local optimum than the other rule bases do. Corresponding to this, if some Levenberg-Marquardt iterations are executed after each bacterial mutational step, the test step is able to choose some potentially valued clones that could be lost otherwise.

In the *Bacterial Memetic Algorithm with the Modified Operator Execution Order*, after *each mutational step* of *every single bacterial mutation iteration* several LM iterations are done. Several tests have shown it is enough to run just 3 to 5 of LM iterations per mutation to improve the performance of the whole algorithm. The usual test phase of the *bacterial mutation operation* follows after the LM iterations, and then, after the complete bacterial mutation follows the LM method that is used in the original BMA, where more, e.g. 10 iterational steps, are done with all the individuals of the population towards reaching of the local optimum. After all this the *gene transfer operation* is done.

The flowchart of the *Bacterial Memetic Algorithm with the Modified Operator Execution Order* can be seen in Fig. 10. In the BMAM method the order of the steps is as follows:

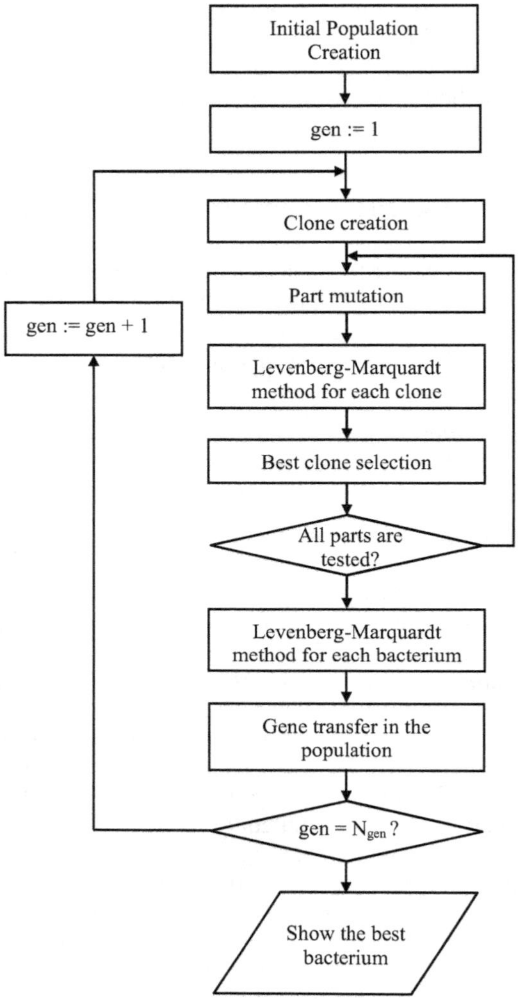


**Fig. 10.** Flowchart of BMAM

1. Apply the *modified bacterial mutation operation* for each individual:
   - Each individual is selected one by one.
   - $N_{Clones}$ copies of the selected individual are created ("clones").
   - Choose the same part or parts randomly from the clones and mutate it (except one single clone that remains unchanged during this mutation cycle).
   - *Run some Levenberg-Marquardt iterations (3 – 5).*
   - *Select the best clone and transfer its all parts to the other clones.*
   - Repeat the part choosing-mutation-*LM*-selection-transfer cycle until all the parts are mutated, *improved* and tested.

- The best individual is remaining in the population, all other clones are deleted.
- This process is repeated until all the individuals have gone through the *modified bacterial mutation.*

2. *Levenberg-Marquardt* method for each individual,
3. *Gene transfer operation* for a partial population.

## 8 Modified Bacterial Memetic Algorithm (MBMA)

Although IBMA and BMAM perform better than the original BMA they behave in different manner in different circumstances. IBMA performed better in case of more complex fuzzy rule base while BMAM performed better in case of less complex fuzzy rule base.

Our recent work has pointed out that combining the improvements in IBMA and BMAM is beneficial (Gál, Botzheim and Kóczy, 2008) [7]. We presented a novel, improved version of the *Bacterial Memetic Algorithm* used for fuzzy rule base extraction named *Modified Bacterial Memetic Algorithm.* We modified the original BMA in two parts. The first one is the *knot order violation handling* concerning the Levenberg-Marquardt method incorporated into the BMA, while the second one is the *operator execution order.*

Although all the parts of the MBMA were described in the previous sections we give the detailed steps of the algorithm that contains all the improvements below (Fig. 11):

1. Create the initial population: $N_{Ind}$ individuals are randomly created and evaluated. ($N_{Ind}$ is the number of individuals in the population.) Each individual contains $N_{Fuzzy\_rules}$ fuzzy rules encoded in the chromosome ($N_{Fuzzy\_rules}$ is the number of fuzzy rules of the desired model).
2. Apply the *Modified Bacterial Mutation operation* for each individual:
    - Each individual is selected one by one.
    - $N_{Clones}$ copies of the selected individual are created ("clones").
    - Choose the same part or parts randomly from the clones and mutate it (except one single clone that remains unchanged during this mutation cycle).
    - *Run some Levenberg-Marquardt iterations (3–5)*
        - *Use method Swap for handling the knot order violations after each LM update.*
    - *Select the best clone and transfer its all parts to the other clones.*
    - Repeat the part choosing-mutation-*LM*-selection-transfer cycle until all the parts are mutated, *improved* and tested.
    - The best individual is remaining in the population, all other clones are deleted.
    - This process is repeated until all the individuals have gone through the *modified bacterial mutation.*
3. Apply the *Levenberg-Marquardt* method to each individual (e.g. 10 iterations per individual per generation).

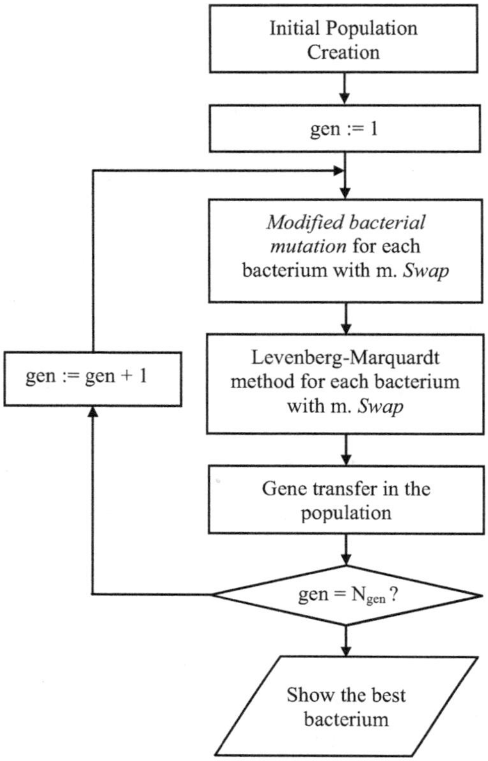


**Fig. 11.** Flowchart of MBMA

- *Use method Swap for handling the knot order violations after each LM update.*

4. Apply the *gene transfer operation* $N_{Inf}$ times per generation:
   - Sort the population according to the fitness values and divide it in two halves. The half that contains the better individuals is called superior half while the other half is the inferior half.
   - Choose one individual (the "source chromosome") from the superior half and another one (the "destination chromosome") from the inferior half.
   - Transfer a part from the source chromosome to the destination chromosome (select the part randomly or by a predefined criterion).
   - Repeat the steps above $N_{Inf}$ times ($N_{Inf}$ is the number of "infections" in one generation.)
5. Repeat the procedure above from the *modified bacterial mutation* step until a certain termination criterion is satisfied (e.g. maximum number of generations).

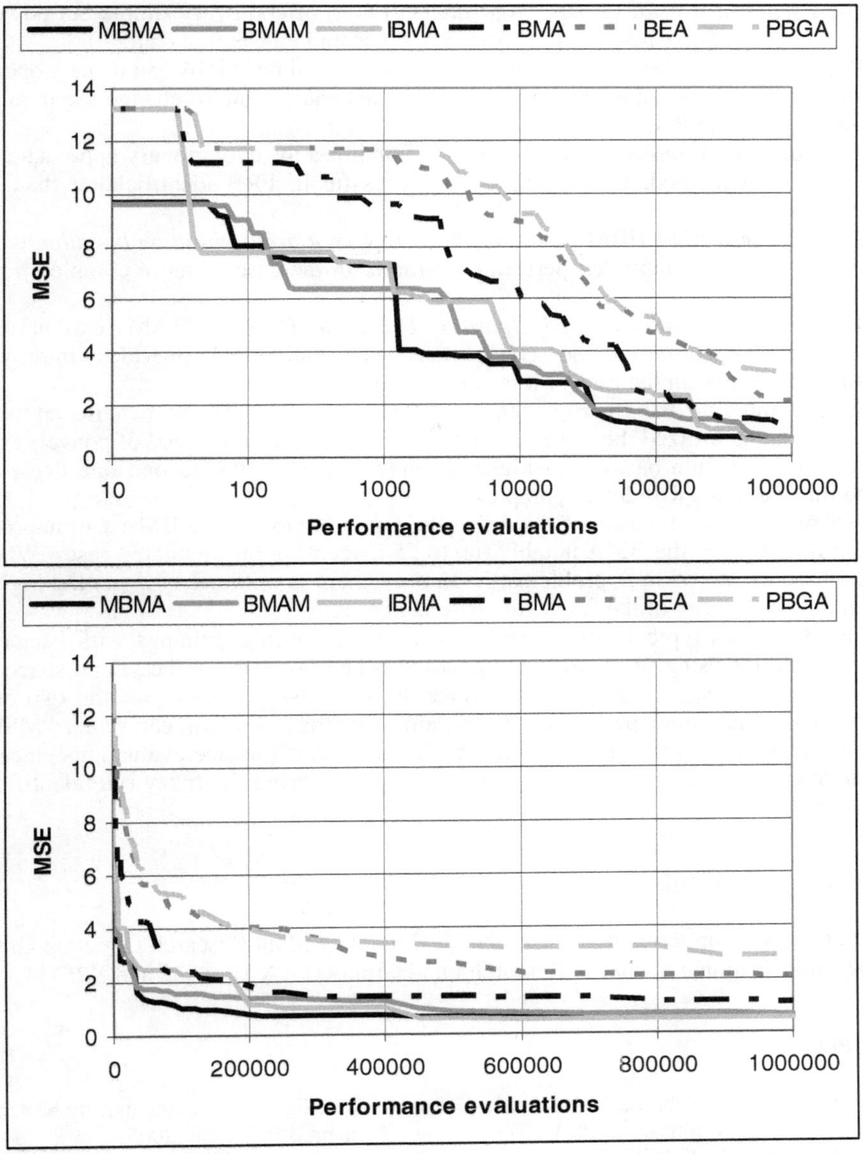


**Fig. 12.** Performance of the bacterial type algorithms used for FRBI

## 9 Conclusions

In this work we summarized the *bacterial type* algorithms used for *fuzzy rule base identification*.

The *Pseudo-Bacterial Genetic Algorithm* (PBGA) offers a very simple but powerful way to extract quasi-optimal fuzzy rule bases from input-output data.

The *Bacterial Evolutionary Algorithm* (BEA) is based on PBGA and its new operator establishes relationships among the individuals and is able to change the number of the rules in the FRB.

The *Bacterial Memetic Algorithm* (BMA) combines the evolutionary approach and a local search method. It is much more successful in FRB identification than its predecessors.

The *Improved BMA* (IBMA) has an alternative *knot order violation handling* technique and provides improved performance rather in the case of more complex *fuzzy rule base*.

The *BMA with the Modified Operator Execution Order* (BMAM) exploits the *Levenberg-Marquardt* method (LM) more efficiently and provides improved performance rather in the case of less complex *fuzzy rule base*.

With combining the improvements of IBMA and BMAM the benefits of both methods can be utilized, because the first method increases the speed of convergence rather for fuzzy rule bases with higher complexity, while the second one does the same for rule bases with lower complexity.

Our previous work has confirmed that the latest version of the BMA can improve the performance of the BMA notably (up to 55 percent) in the simulated cases. While in the case of very simple problems the improvement is minimal, it is getting higher when the complexity of the fuzzy rule base increases.

Fig. 12 shows typical differences in performance during trainings with bacterial type algorithms using identical initial conditions and positions in the search space (2 input variables, $N_{Patterns}$=200, 3 fuzzy rules, $N_{Ind}$=10, $N_{Clones}$=8, $N_{Inf}$=4, the two diagrams show the same performance diagram with logarithmic/linear scale, "MSE" means the Mean Squared Errors of the model, and "Performance evaluations" means the number of model performance evaluations made during the fuzzy rule identification process.)

## Acknowledgments

This paper was supported by the Széchenyi University Main Research Direction Grant 2009, and a National Scientific Research Fund Grant OTKA T048832 and K75711.

## References

[1] Botzheim, J., Cabrita, C., Kóczy, L.T., Ruano, A.E.: Fuzzy rule extraction by bacterial memetic algorithms. In: IFSA 2005, Beijing, China, pp. 1563–1568 (2005)

[2] Botzheim, J., Cabrita, C., Kóczy, L.T., Ruano, A.E.: Estimating Fuzzy Membership Functions Parameters by the Levenberg-Marquardt Algorithm. In: FUZZ-IEEE 2004, Budapest, Hungary, pp. 1667–1672 (2004)

[3] Botzheim, J., Kóczy, L.T., Ruano, A.E.: Extension of the Levenberg-Marquardt algorithm for the extraction of trapezoidal and general piecewise linear fuzzy rules. In: IEEE World Congress on Computational Intelligence, Honolulu, pp. 815–819 (2002)

[4] Cabrita, C., Botzheim, J., Gedeon, T.D., Ruano, A.E., Kóczy, L.T., Fonseca, C.: Bacterial Memetic Algorithm for Fuzzy Rule Base Optimization. In: World Automation Congress, WAC 2006 (2006)

[5] Gál, L., Botzheim, J., Kóczy, L.T., Ruano, A.E.: Fuzzy Rule Base Extraction by the Improved Bacterial Memetic Algorithm. In: 6th International Symposium on Applied Machine Intelligence and Informatics, Herl'any, Slovakia, January 21-22, pp. 49–53 (2008)

[6] Gál, L., Botzheim, J., Kóczy, L.T.: Improvements to the Bacterial Memetic Algorithm used for Fuzzy Rule Base Extraction. In: Computational Intelligence for Measurement Systems and Applications, CIMSA 2008, Istanbul, Turkey, pp. 38–43 (2008)

[7] Gál, L., Botzheim, J., Kóczy, L.T.: Modified Bacterial Memetic Algorithm used for Fuzzy Rule Base Extraction. In: 5th International Conference on Soft Computing as Transdisciplinary Science and Technology, CSTST 2008, Paris, France, pp. 425–431 (2008)

[8] Holland, J.H.: Adaptation in Nature and Artificial Systems: An Introductory Analysis with Applications to Biology, Control, and Artificial Intelligence. The MIT Press, Cambridge (1992)

[9] Johanyák, Z.C.: Fuzzy rule interpolation methods and automatic system generation based on sample data (in Hungarian), Ph.D thesis, University of Miskolc (2007)

[10] Johanyák, Z.C., Kovács, S.: Sparse Fuzzy System Generation by Rule Base Extension. In: Proceedings of the 11th IEEE International Conference of Intelligent Engineering Systems (IEEE INES 2007), Budapest, Hungary, pp. 99–104 (2007)

[11] Johanyák, Z.C., Kovács, S.: Fuzzy Rule Interpolation Based on Polar Cuts. In: Computational Intelligence, Theory and Applications, pp. 499–511. Springer, Heidelberg (2006)

[12] Johanyák, Z.C., Tikk, D., Kovács, S., Wong, K.W.: Fuzzy Rule Interpolation Matlab Toolbox - FRI Toolbox. In: Proceedings of the IEEE World Congress on Computational Intelligence (WCCI 2006), 15th Int. Conf. on Fuzzy Systems (FUZZ-IEEE 2006), Vancouver, BC, Canada, pp. 1427–1433. Omnipress (2006) ISBN 0-7803-9489-5

[13] Klawonn, F., Kruse, R.: Constructing a fuzzy controller from data. Fuzzy Sets and Systems 85(2), 177–193 (1997)

[14] Kóczy, L.T., Hirota, K.: Size reduction by interpolation in fuzzy rule bases. IEEE Transactions on System, Man and Cybernetics 27, 14–25 (1997)

[15] Kovács, S.: Extending the Fuzzy Rule Interpolation "FIVE" by Fuzzy Observation. In: Reusch, B. (ed.) Advances in Soft Computing, Computational Intelligence, Theory and Applications, pp. 485–497. Springer, Germany (2006) ISBN 3-540-34780-1

[16] Kovács, S., Kóczy, L.T.: The use of the concept of vague environment in approximate fuzzy reasoning. In: Fuzzy Set Theory and Applications, Bratislava, Slovakia, vol. 12, pp. 169–181. Tatra Mountains Mathematical Publications, Mathematical Institute Slovak Academy of Sciences (1997)

[17] Mamdani, E.H., Assilian, S.: An experiment in linguistic synthesis with a fuzzy logic controller. Int. J. Man-Mach. Stud. 7, 1–13 (1975)

[18] Marquardt, D.: An Algorithm for Least-Squares Estimation of Nonlinear Parameters. SIAM J. Appl. Math. 11, 431–441 (1963)

[19] Moscato, P.: On evolution, search, optimization, genetic algorithms and martial arts: Towards memetic algorithms, Technical Report Caltech Concurrent Computation Program, Report. 826, California Institute of Technology, Pasadena, California, USA (1989)

[20] Nawa, N.E., Hashiyama, T., Furuhashi, T., Uchikawa, Y.: A study on fuzzy rules discovery using pseudo-bacterial genetic algorithm with adaptive operator. In: Proceedings of IEEE Int. Conf. on Evolutionary Computation, ICEC 1997 (1997)

[21] Nawa, N.E., Furuhashi, T.: Fuzzy Systems Parameters Discovery by Bacterial Evolutionary Algorithms. IEEE Transactions on Fuzzy Systems 7, 608–616 (1999)

[22] Sugeno, M., Kang, K.T.: Structure Identification of Fuzzy Model. Fuzzy Sets and Systems 28, 15–33 (1988)

[23] Brand, M.: Fast Low-Rank Modifications of the Thin Singular Value Decomposition. Linear Algebra and Its Applications 415(1), 20–30 (2006)

# 4

# Discovering Associations with Uncertainty from Large Databases*

Guoqing Chen**, Peng Yan, and Qiang Wei

Department of Management Science and Engineering
School of Economics and Management, Tsinghua University,
Beijing 100084, China

**Abstract.** Data mining, also known as knowledge discovery in databases, is the process of extracting desirable knowledge or interesting patterns from existing databases. As a specific form of knowledge, association reflects semantics in terms of relationships among attributes in databases, and has been widely studied recently. This chapter focuses on dealing with uncertainty in discovering association rules (AR) and functional dependencies (FD), and provides an overview of our efforts on association rules with fuzzy taxonomies (FAR), on implication-based fuzzy quantitative association rules (ARsi), and on functional dependencies with partial degrees of satisfaction ($FD_d$).

## 1 Introduction

### 1.1 Mining Associations in Databases

The past few decades have witnessed an explosion in the amount of electronically stored data due to advances in information technology and massive applications in business and scientific domains. Data mining, sometimes also referred to as knowledge discovery in databases (KDD), is regarded as a non-trivial process of identifying valid, novel, potentially useful, and ultimately understandable knowledge in large-scale data. The scope of data mining and KDD is broad and can be viewed as a multitude of fields of study related to statistics theories, machine learning, artificial intelligence, fuzzy logic, database theories, image processing, pattern recognition, data visualization, etc. Generally, the task of data mining may concentrate on clustering, classification, association, prediction, regression, summarization, time-series analysis and deviation detection, and so on. Data mining techniques have been widely used in many applications and fields such as marketing, stock and finance, geography, aerography, engineering and economics. (Fayyad & Piatetsky-Shapiro et al., 1996; Han J. and Kamber, 2000) As a specific kind of knowledge, association that reflects relationships among attributes of databases is discussed in this chapter. Association rules and functional dependencies are of particular concern.

Association rules, proposed by Agrawal & Imielinski et al. in 1993, have become a focal point of research and applications in data mining. They were first used to solve the

* The work was partly supported by the National Natural Science Foundation of China (70621061/70890080), and Tsinghua's Research Center for Contemporary Management.

** Corresponding author, gqchen@tsinghua.edu.cn

E. Rakus-Andersson et al. (Eds.): Recent Advances in Decision Making, SCI 222, pp. 45–66.
springerlink.com 

problems (also called "market-basket problems") in supermarkets to improve the layout of goods in stores. Typically, an association rule is of the form X⇒Y and expresses the semantics that occurrences of Y are associated with occurrences of X. Table 1 shows an example of a purchase relation in a supermarket. An association rule "Pork ⇒Tomato" means that "a customer who buys Pork is likely to buy Tomato as well".

**Table 1.** The purchase relation in a supermarket

| *transid* | *Custid* | *Goods* |
|---|---|---|
| 111 | 201 | Tomato, Pork, Cabbage |
| 112 | 105 | Sausage, Apple |
| 113 | 311 | Tomato, Pork, Apple |
| 114 | 201 | Tomato, Cabbage, Apple |
| 115 | 118 | Tomato, Pork, Cabbage, Apple |

In table 1, each value of attribute Goods can be viewed as an item. Then the purchase database can be transferred into a binary databases (table 2), in which values of each item are 0's or 1's. Association rules associating such binary-valued attributes are usually referred to as Boolean association rules (or simply, association rules, otherwise indicated where necessary).

**Table 2.** Example of database D converted from the purchase relation

| *D* | *Tomato* | *Pork* | *Apple* | *Cabbage* | *Sausage* |
|---|---|---|---|---|---|
| 111 | 1 | 1 | 0 | 1 | 0 |
| 112 | 0 | 0 | 1 | 0 | 1 |
| 113 | 1 | 1 | 1 | 0 | 0 |
| 114 | 1 | 0 | 1 | 1 | 0 |
| 115 | 1 | 1 | 1 | 1 | 0 |

Formally, let $I = \{I_1, I_2 \ldots I_m\}$ be a set of items, D be a database of n tuples (or records) with respect to relation R(I), in which each tuple t is represented as a subset of I, $t[I_k] \in \{0, 1\}$ $(k = 1, 2, \ldots, m)$, and X be a set of items (also referred to as an itemset), $X \subset I$, then t is called to support X if for all items $I_k \in X$, $t[I_k] = 1$. X is called a k-itemset if X contains k items. The degree of support (Dsupport) for itemset X is the percentage of tuples containing X in the database concerned, and is defined as: $Dsupport(X) = \|X\|/|D|$ (Agrawal & Imielinski et al., 1993; Agrawal & Srikant, 1994; Agrawal & Mannila et al., 1996). As an example in Table 2, Dsupport(Pork) = 3/5 = 60% because the item Pork appears in 3 transactions.

For association rule X⇒Y, X and Y are two disjoint itemsets of I, i.e., X, Y⊂I, X∩Y = ∅. Two measures, namely, degree of support (Dsupport) and degree of confidence (Dconfidence), are used to evaluate the rule's significance and strength respectively. Dsupport(X⇒Y) is the percentage of tuples containing both X and Y in the database, i.e., Dsupport(X⇒Y) = Dsupport(X∪Y) = ||X∪Y||/|D|, and Dconfidence(X⇒Y) is the ratio of the number of tuples that contain both X and Y over the number of tuples that contain X, i.e., Dconfidence(X⇒Y) = ||X∪Y||/||X||, where ||X|| is the number of tuples in D that support X, ||X∪Y|| is the number of tuples in D that support the itemset X∪Y, and |D| is the number of tuples in D. Given two basic thresholds: minimal support $\alpha$ and minimal confidence $\beta$, itemset X is called a frequent itemset if Dsupport(X) $\geq \alpha$. Rule X⇒Y is called a valid association rule if Dsupport(X⇒Y) $\geq \alpha$ and Dconfidence(X⇒Y) $\geq \beta$. Statistically, Dsupport could be regarded as the significance of a rule supported by a database, while Dconfidence could be regarded as the certainty of a rule. For the database in table 2, for example, rule Pork⇒Tomato's Dsupport and Dconfidence are: Dsupport(Tomato⇒Pork) = Dsupport(Tomato∪Pork) = 3/5 = 60%, Dconfidence (Tomato⇒Pork) = Dsupport(Tomato⇒Pork)/ Dsupport(Tomato) = 3/4 = 75%.

The algorithm proposed by Agrawal & Srikant et al. (1994) to discover association rules is considered as a typical mining algorithm, namely the Apriori algorithm. The algorithm consists of two steps: calculating all frequent itemsets with the user-specified minimum support, and generating all rules with minimum confidence using frequent itemsets as input.

Generally, the step of generating frequent itemsets is more important and time consuming, and attracts more research efforts. The main idea of generating frequent itemsets is to first generate candidate ($k$+1)-itemsets based on frequent $k$-itemsets by *join operation* and *pruning strategy*, and then to calculate frequent ($k$+1)-itemsets based on candidate ($k$+1)-itemsets by scanning databases. The algorithm proceeds in this way until the set of frequent $k$-itemsets or the set of candidate ($k$+1)-itemsets is null. *Join operation* is that any two frequent $k$-itemsets with same ($k$-1)th items and different $k$th items can be joined into a new ($k$+1)-itemset using union operation. For example, given $\alpha$=50%, frequent 2-itemsets {Tomato, Pork} (Dsupport=60%) and {Tomato, Apple} (Dsupport=60%) have the same first items (i.e., Tomato) and different second items (i.e., Pork and Apple, respectively), we may get a 3-itemset {Tomato, Pork, Apple} resulting from {Tomato, Pork}∪{Tomato, Apple}. Furthermore, the *pruning strategy* is based on the property that any subset of a frequent itemset must be frequent, which can be incorporated into the algorithm to greatly reduce the number of candidate itemsets to generate, and hence to improve the algorithm's efficiency. In our example, itemset {Tomato, Pork, Apple} will not be inserted into the candidate set because its subset {Pork, Apple} is not frequent (with Dsupport=40%).

The investigation of association rules can be categorized in two directions. One is to improve the algorithms' computatinal efficiency as discussed in (Houtsma & Swarmi, 1993; Fayyad & Uthurusamy, 1994; Mannila & Toivonen, 1994; Savasere & Omiecinski et al., 1995; Agrawal & Mannila, 1996). In addition, some methods also construct their algorithms upon sampling operations (Yilmaz & Triantaphyllou et al., 2002). In addition to the above serial algorithms, some parallel and distributed algorithms are also presented (Mueller, 1995; Agrawal & Shafer 1996). The other direction is to extend the semantics and expressions of rules from a number of perspectives.

Srikant & Agrawal (1995) and Han & Fu (1995) presented methods to discover generalized association rules (GARs), by which rules between different levels of taxonomies can be obtained. Thus, a GAR reflects semantics that not only items but also their superclasses are associated with each other. Moreover, according to types of the domains of the attributes, quantitative/categorical association rules (Srikant & Agrawal 1996) and association rules to deal with time series and temporal data are investigated (Chen & Ai et al., 2002, Chen & Wei et al., 2001, Zhang & Chen et al., 2008). Furthermore, some other studies focused on improving the interestingness of the discovered association rules, such as the association rules with constraints and contexts (Fukuda & Morimoto et al, 1996; Han & Fu, 1995; Klemettinen & Mannila et al., 1994; Srikant & Vu et al., 1997). Apart from Dsupport and Dconfidence measures, some other interestingness measures, based on statistics and information theory, have also been proposed aimed at making the discovered rules more understandable or simpler (Tseng, 2001; Maimon & Kandel et al., 2001, Chen & Wei et al., 2002).

In addition, functional dependency (FD) can be viewed as another important kind of association of interest. Notable, FD is a major issue in the design of relational databases, and is considered as a piece of semantic knowledge in terms of integrity constraints, which is used to reduce update anomalies in databases (Codd, 1970; Chen, 1998). For attribute collections X and Y in a relational schema of database D, a FD, X→Y, represents the semantics that equal Y-values correspond to equal X-values. More concretely, X→Y ⇔ for any two tuples t and t' in D, if t[X] = t'[X] then t[Y] = t'[Y]. An example of FD is "equal *tranid* leads to equal *cusid*".

Classically, functional dependencies could be assumed or constructed logically, based on which relation schemas are designed. In the context of data mining as a type of reverse engineering, the discovery of functional dependencies has received considerable attentions recently (Castellanos & Saltor, 1993; Huhtala & Karkkainen, 1998a, 1998b; Liao & Wang et al., 1999; Savnik & Flach, 2000; Bosc & Pivert et al., 2001; Wei & Chen et al., 2002). The basic idea behind is that numerous database applications over decades have generated and maintained a huge amount of data stored in distributed environments and with diversified structures. Many functional dependencies might not originally be known or thought of being important, or have been hidden over time, but may be useful and interesting as integrity constraints and semantic knowledge.

## 1.2 Fuzziness in Association Mining

In many situations, the process of discovering the above-mentioned associations involves uncertainty, and treatment of uncertainty is considered as one of the key issues in data mining (Fayyad & Uthurusamy, 1994; Kruse & Nanck et al., 2001; Rifqi & Monties, 2001). Actually, typical association rules are discovered from the viewpoint of probabilities. Dsupport of an association rule X⇒Y could be regarded as the estimation of Pr(XY), while Dconfidence of X⇒Y as the estimation of Pr(Y|X) (Aumann & Lindell, 1999).

Fuzziness is another very important type of uncertainty, and has been widely introduced into the field of knowledge representations and data mining. It is necessary and important to apply fuzzy logic to data mining for two reasons: one is that fuzziness is inherent in many problems of knowledge representation and discovery, and the other is that high-level managers or complex decision processes often deal with generalized concepts and linguistic expressions, which are generally fuzzy in nature. Fuzzy logic, or interchangeably referred to as fuzzy sets theory, had its inception by Zadeh (1965), and plays an important role in dealing with fuzziness and therefore in fuzzy data mining.

In association rules mining, the data items concerned may be categorized in classes upon specific properties, which can be represented in hierarchies or taxonomies in terms of subclass and superclass (e.g., apple, fruit, food, etc.), data mining may refer to data items at different levels of taxonomies. For instance, GAR deals with the relationships across taxonomic nodes of higher levels, reflecting more general semantics, such as "Fruit ⇒ Meat" instead of "Apple ⇒ Pork". However, there are situations where a subclass belongs to its superclass at a partial degree in [0, 1], resulting in fuzzy taxonomies. For example, the item tomato may be regarded to belong to both fruit and vegetable at 0.7 and 0.6 respectively.

Furthermore, quantitative association rules are considered important and meaningful because real applications always contain data with quantitative values but not restrict to {0, 1}. The traditional method proceeded by partitioning attribute domains into several intervals and transforming the quantitative values into binary ones in order to apply the classical mining algorithm, which will induce the "sharp boundary" problem. For example, consider rules like "if the customers are at ages of [20, 30], then they tend to buy electronics at price of [$5000, $10000]". Apparently, a customer aged 31 with a purchase of $8000 may not be identified. An alternative expression of the rule may be "Young customers tend to buy expensive electronics". This expression is more flexible and more general in semantics and could reflect this customer's buying behavior in a natural way. Notably, here "young customers" and "expensive electronics" are linguistic terms that are fuzzy in nature.

Since the concept of fuzziness can be incorporated into the model of association rules, related classical operations need to be extended with fuzzy set operations. Concretely, intersection, union and implication operations on fuzzy sets are usually relevant, which may be defined by means of t-norms, t-contorms and implication operators respectively. T-norms, t-contorms and fuzzy implication operators are all $[0, 1]^2 \rightarrow [0, 1]$ mappings. Desirably, a t-norm T is commutative, associative and satisfies $T(a, 1) = a$, $T(a, 0) = 0$ for every $a \in [0, 1]$; a t-conorm S is commutative, associative and satisfies $S(a, 1) = 1$ and $S(a, 0) = a$ for every $a \in [0, 1]$; and a fuzzy implication operator (FIO) I is decreasing in its first component and increasing in its second component, and satisfies $I(0, 0) = (0, 1) = I(1, 1) = 1$ and $I(1, 0) = 0$. The problem of which operators to choose depends specifically on the application at hand and the properties that need to be fulfilled. Some well-know t-norms, t-conorm and implication operators are listed as follows (Klir & Yuan, 1999):

T-norm: $T(a, b) = \min(a, b)$, $T(a, b) = ab$, $T(a, b) = \max(a+b-1,0)$ ;
T-conorm: $S(a, b) = max(a, b)$, $S(a, b) = a+b-ab$, $S(a, b) = \min(a+b, 1)$;
Fuzzy Implication Operators: $I(a, b) = \min(1, 1 - a + b)$, $I(a, b) = 1 - a + ab$, $I(a, b) = \max(1-a,b)$, $I(a,b) = \begin{cases} 1 \text{ } if \text{ } a \leq b \\ b, otherwise \end{cases}$

Moreover, the concept of uncertainty can also be used in the model of mining functional dependencies. In real applications, null values, noise data and conflicts do always exist and it is necessary to allow some exceptions to enhance the robust of the mining methods. Therefore, we may need to express functional dependency "X functionally determines Y" in a more tolerable manner as a general setting that it holds in a partial degree.

### 1.3 Related Work

Fuzzy association rules with fuzzy taxonomies and FAR with linguistic hedges are introduced in (Chen & Wei, et al., 1999, Chen & Wei, 2002). To deal with the "sharp boundary" problem in partitioning quantitative data domains, fuzzy quantitative association rules are discussed in (Fu et al., 1998; Mazlack, 2000; Chien & Lin et al., 2001; Graff et al., 2001; Gyenesei, 2000a; Hong & Kuo et al., 1999a, 1999b). From a more logic oriented viewpoint, the notion of fuzzy implication is also incorporated into the fuzzy association rules in (Chen & Yan et al., 2004, Yan & Chen 2005, Hullermeier, 2001a, 2001b). Different fuzziness-related interestingness measures are proposed in (Wei & Chen, 1999, Kuok & Fu et al., 1998, Gyenesei & Teuhola, 2001, Au & Chan 1997, 1998) to extend the frame of Dsupport-Dconfidence. Weighted association rules (Cai & Fu et al., 1998, Gyenesei, 2000b, Shu & Tsang et al., 2000) are applied to distinguish the importance of different items. Other research on fuzzy extensions on association rules mining can be seen in (Luo, 1999, 2000, De Cock & Cornelis et al., 2003).

Mining functional dependencies are attracting more and more attentions to enrich knowledge of designing databases. Cubero et al. (1995, 1999) presented a method of data summarization through fuzzy functional dependencies in both crisp and fuzzy databases, in which projection operations are applied to reduce the amount of data in databases without loss of information. Wang & Shen et al. (2002) presented a method to discover fuzzy functional dependencies in similarity-based relational databases with an incremental strategy, which has advantage in dealing with non-static databases. Huhtala et al. (1998a, 1998b) explored approximate dependency so as to represent functional dependency which "almost holds". Wei & Chen (2002) introduce functional dependencies with degrees ($FD_d$) to tolerate noises or null data which exist mostly in real databases. And Yang & Singhal 2001 attempted at presenting a framework of linking fuzzy functional dependencies and fuzzy association rules in a closer manner.

The remaining part of the chapter will concentrate on an overview of our efforts on association rules in fuzzy taxonomies (Chen & Wei, 2002), association rules in quantitative databases from the viewpoint of fuzzy implication (Chen, Yan & Kerre, 2004; Yan & Chen, 2005), and functional dependencies with degrees of satisfaction (Wei & Chen, 2004).

## 2 Fuzzy Logic in Association Rules

### 2.1 Fuzzy Association Rules with Fuzzy Taxonomies

#### 2.1.1 Generalized Association Rules

As mentioned previously, Srikant & Agrawal (1995) and Han & Fu (1995) presented methods to discover association rules between difference levels of concept taxonomies for two reasons. One is that the model provides one kind of more meaningful knowledge for users or managers. And the other is that it avoids the problem that some strong rules with low support items in leaf nodes may be ignored. The proposed algorithms in Srikant & Agrawal 1995 and Han & Fu 1995 allows the discovery of association rules that represent the relationships between original items, as well as between items at all levels of related taxonomies, e.g., "Fruit⇒Meat", which is more general and have more potential to be discovered. As shown in Figure 1 (a), it is called generalized association rules.

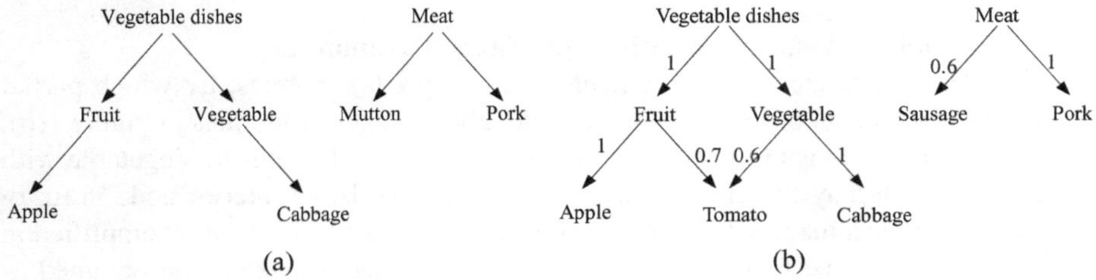


**Fig. 1.** Exact Taxonomies and Fuzzy Taxonomies

Formally, generalized association rules could be illustrated as follows. For original $I = \{I_1, I_2, \ldots, I_m\}$ and databases D with respect to relation R(I), $\forall t \in D$, $t[I_k]$ belongs to {0 ,1}, k = 1, 2, …, m. Given a collection G of taxonomies, in which all leaf items belong to I, then adding all the interior items (nodes) of G into I will result in a new set of items $I_G$. Subsequently, a new database $D_G$ with respect to $R(I_G)$ can be derived, in which each tuple t in $D_G$ is also a binary vector. For any $I_k \in I_G$, if $I_k \in I$, then $t[I_k] = t[I_k]$. If $I_k \in (I_G - I)$, then (i) $t[I_k] = 1$ if there exists any descendant $I_k'$ of $I_k$ that $t[I_k'] = 1$; (ii) $t[I_k] = 0$ otherwise. Likewise, let itemset X be a subset of $I_G$, then a tuple t is called to support X if for any item $I_k \in X$, $t[I_k] = 1$. Thus, mining generalized association rules in D on R(I) with taxonomies G becomes to discover Boolean association rules in $D_G$ on $R(I_G)$.

As an example, Table 3 shows a database $D_G$ with respect to relation R(Tomato, Pork, Apple, Cabbage, Sausage, Fruit, Vegetable, Vegetable-dishes, Meat) in accordance with G in Figure 1(a) and purchase relation in table 1. Notably, $D_G$ degenerates to D when projecting R(Tomato, Pork, Apple, Cabbage, Sausage, Fruit, Vegetable, Vegetable-dishes, Meat) on (Tomato, Pork, Apple, Cabbage, Sausage).

**Table 3.** Example of extended database $D_G$ in accordance with G in Figure 1(a) and table 1

| $D_G$ | Tomato | Pork | Apple | Cabbage | Sausage | Fruit | Vegetable | Vegetable-dishes | Meat |
|---|---|---|---|---|---|---|---|---|---|
| 111 | 1 | 1 | 0 | 1 | 0 | 0 | 1 | 1 | 1 |
| 112 | 0 | 0 | 1 | 0 | 1 | 1 | 0 | 1 | 0 |
| 113 | 1 | 1 | 1 | 0 | 0 | 1 | 0 | 1 | 1 |
| 114 | 1 | 0 | 1 | 1 | 0 | 1 | 1 | 1 | 0 |
| 115 | 1 | 1 | 1 | 1 | 0 | 1 | 1 | 1 | 1 |

In dealing with general association rules in taxonomies, two optimization strategies can be used in the mining process. One is that Dsupport for an itemset X that contains both an item x and its ancestor x^ will be the same as the support for the itemset X-x^, then join operation will avoid the problem of both the item and it's ancestor in the same itemsets. The other is that Dsupport for itemsets in low levels is no more than that in high levels, which can be used in the algorithm as a pruning strategy, but there exists a tradeoff between this strategy and times of scanning databases. Furthermore, one interestingness measure, namely Interest, is introduced to remove some non-interesting rules from the viewpoint of probability expectation.

### 2.1.2 Generalized Association Rules with Fuzzy Taxonomies

We extended generalized association rules with fuzzy taxonomies, by which partial belongings could be incorporated. For example, given fuzzy taxonomies in Figure 1(b), Tomato not only belongs to Fruit with degree 0.7, but also belongs to Vegetable with degree 0.6, which may be semantically meaningful. That is, an interior node in fuzzy taxonomies can be a fuzzy set. Generally, given fuzzy taxonomies $G^f$ as exemplified in Figure 1(b), the degree that any node y belongs to its ancestor x can be obtained as follows:

$$\mu_{xy} = \underset{\forall l: x \to y}{S} \left( \underset{\forall e\ on\ l}{T} \mu_{le} \right)$$

where *l: x→y* is one of the accesses (paths) of attributes *x* and *y*, *e on l* is one of the edges on access *l*, $\mu_{le}$ is the degree on the edge *e* on *l*. If there is no access between *x* and *y*, $\mu_{xy} = 0$. S is t-conorm and T is t-norm. In Chen & Wei 2002, max for S and min for T are employed, others are also available. Then based on all the $\mu_{xy}$ derived between any two nodes, an interior item in $G^f$ could be represented as a fuzzy set, each element of which is a leaf item with its membership degree to the interior item. For example, in Figure 1(b), the items Fruit, Vegetable, Vegetable dishes and Meat are all fuzzy sets, and Fruit = {1/Apple, 0.7/Tomato}, Vegetable = {0.6/Tomato, 1/Cabbage}, Vegetable dishes = {1/Apple, 0.7/Tomato, 1/Cabbage}, Meat = {0.6/Sausage, 1/Pork}.

Then, with original I, D, and given $G^f$, the newly obtained set of items $I_{Gf}$ is in the same way as $I_G$ discussed in section 2.1, except for the fact that any interior item in $I_{Gf}$ is generally a fuzzy set, not an ordinary super-class. Correspondingly, the extended database $D_{Gf}$ can be derived from D on R(I) such that $\forall t \in D_{Gf}$, $\forall I_k \in I_{Gf}$, $t[I_k] = \max_{\forall L \in I}(\mu_{I_k L})$. $D_{Gf}$ in accordance with Figure 1(b) and Table 3 is tabulated in Table 4, if S(*a*, *b*) = max(*a*, *b*) and T(*a*, *b*) = min(*a*, *b*), for example, 111[Fruit]= max($\mu_{FruitApple}$, $\mu_{FruitTomato}$) = max(min(0, 1), min(1, 0.7)) = 0.7.

In addition, let X be a fuzzy itemset in $I_{Gf}$, then a tuple t in $D_{Gf}$ is called to support X with a certain degree $t[X] = \underset{I_k \in X}{T} t[I_k]$. Furthermore, an association rule in fuzzy taxonomies is of the form: X⇒Y, where X and Y are fuzzy itemsets, X, Y∈ $I_{Gf}$ and X∩Y = ∅. The degree of support for X is extended as follows:

$$\text{Dsupport}(X) = \frac{\|X\|}{|D_{Gf}|} = \frac{\sum count_{t \in D_{Gf}} (\underset{I_k \in X}{T} t[I_k])}{|D_{Gf}|} = \frac{\sum count_{t \in D_{Gf}} (t[X])}{|D_{Gf}|}$$

**Table 4.** Example of extended database $D_{Gf}$ in accordance with $G^f$ in Figure 2(b)

| $D_{Gf}$ | Tomato | Pork | Apple | Cabbage | Sausage | Fruit | Vegetable | Vegetable-dishes | Meat |
|---|---|---|---|---|---|---|---|---|---|
| 111 | 1 | 1 | 0 | 1 | 0 | 0.7 | 1 | 1 | 1 |
| 112 | 0 | 0 | 1 | 0 | 1 | 1 | 0 | 1 | 0.6 |
| 113 | 1 | 1 | 1 | 0 | 0 | 1 | 0.6 | 1 | 1 |
| 114 | 1 | 0 | 1 | 1 | 0 | 1 | 1 | 1 | 0 |
| 115 | 1 | 1 | 1 | 1 | 0 | 1 | 1 | 1 | 1 |

where $|D_{Gf}|$ is the number of all tuples in $D_{Gf}$, and $||X||$ is $\sum$count of tuples in $D_{Gf}$ supporting the itemsets X, also called fuzzy cardinality of X. In real applications, different t-norm such as product ($T(a, b) = ab$) and min ($T(a, b)=\min(a, b)$) can be used. For instance, in (Chen & Wei, 2002), min operator is used, while in (Kuok & Fu, 1999; Gyenesei, 2000a), product operator is used, depending on different contexts. Moreover, Dsupport and Dconfidence for rule X⇒Y are straightforward extended as follows:

$$\text{Dsupport}(X \Rightarrow Y) = \text{Dsupport}(X \cup Y) = \frac{\sum count_{t \in D_{Gf}}(t[X \cup Y])}{|D_{Gf}|}$$

$$\text{Dconfidence}(X \Rightarrow Y) = \frac{\text{Dsupport}(X \cup Y)}{\text{Dsupport}(X)} = \frac{\sum count_{t \in D_{Gf}}(t[X \cup Y])}{\sum count_{t \in D_{Gf}} t[X]}$$

As an example in table 4, for the rule Fruit⇒Meat, Dsupport(Fruit⇒Meat) = (min(0.7, 1) + min(0.6, 1) + min(1, 1) + min(1, 0) + min(1, 1))/5 = 3.3/5 = 66%, and Dconfidence(Fruit⇒Meat) = 3.3/(0.7 + 1 + 1 + 1 +1) = 70.2%.

### 2.1.3 Fuzzy Association Rules with Linguistic Hedges

In this section, we will consider the work on linguistic hedge used to modify fuzzy association rules, aimed at generalizing and enriching knowledge representation semantically. We present an approach to incorporate linguistic hedges on existing fuzzy taxonomies to express more fruitful and natural knowledge. The basic idea is that, in the fuzzy taxonomies $G^f$, an interior node could be expressed as a fuzzy set on its child-nodes, the interior node could be modified in forms of hedges with the same child nodes. Then after applying all the proper hedges in a given linguistic pool H onto the items in $G^f$, new fuzzy taxonomies $G^H$ with all modified items could be derived, as shown in Figure 2. In so doing, the problem of mining linguistic association rules with hedges pool H on fuzzy taxonomies $G^f$ could be transferred to mining fuzzy association rules on the new taxonomic structures $G^H$.

In general, let $I_f = \{I_1, I_2, \ldots, I_m\}$ be a set of fuzzy items, each with a membership function $f_k$ (k = 1,2,...,m), and $D_f$ be a database with schema $R(I_f)$ and a pool of hedges H (where assuming that H contains a certain hedge h (e.g., "same") with $\lambda = 1$ such that for any primary linguistic term hw = w). After applying H on $I_f$, then $I_H$ could be derived as follows: $I_H = \{hI_k \mid hI_k$ is a linguistic item modified by h on $I_k$ with membership function $(f_{I_k})^\lambda$, $h \in H$, and $I_k \in I_f\}$. It can be seen that all the original items and the

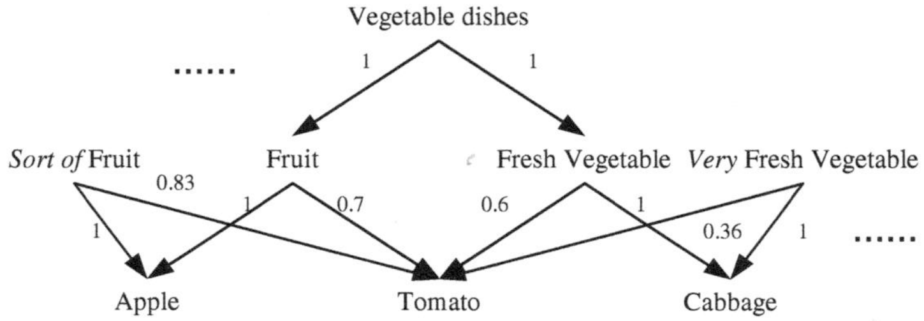


**Fig. 2.** Part of Linguistically Modified Fuzzy Taxonomic Structure

modified items are contained in $I_H$. Moreover, not every h in H can be applied onto $I_k$ in $I_f$. This is due to the semantic constraints of linguistic terms. For example, given $I_f$ = {Young, Fruit}, H = {(Same, 1), (Very, 2), (Sort-of, 1/2)}, then $I_H$ = {Young, Very Young, Sort-of Young, Fruit, Sort-of Fruit}. Further, let $D_H$ is the extended database on schema R($I_H$), in which each tuple t is represented as a vector with $t[hI_k] = [f_{I_k}(t)]^{\lambda}$. After filtering with thresholds α and β given by experts or decision-makers, the discovered rules could look like "Expensive Electronics ⇒ Very Cool Jeans", etc. Generally speaking, this extension of knowledge representation of fuzzy association rules could be represented as rules in forms of $H_X X \Rightarrow H_Y Y$, where X and Y are fuzzy sets and $H_X$ and $H_Y$ are linguistic hedges onto X and Y respectively.

### 2.1.4 R-Interestingness Measure

For association rules with fuzzy taxonomies, degree of interest in generalized association rules can also be extended. For example, consider the taxonomies as shown in Figure 1(b), and suppose that there are 100 transactions containing Fruit and 50 transactions containing Tomato in the database. Since Tomato belongs to Fruit at 0.7, then for a discovered rule Fruit ⇒ Pork (Dsupport = 20%, Dconfidence = 80%), it could be expected readily that Tomato ⇒ Pork has Dsupport of 7% (0.2×(50/100)×0.7) and 80% Dconfidence. If such a rule (Tomato ⇒ Pork at 7% and 80%) is really generated from the database in the mining process, it can be considered redundant since it does not convey any additional information and is less general than the first rule (Fruit ⇒ Pork).

The interesting degree for rule X⇒Y is defined as:

$$\text{Interest}(X \Rightarrow Y) = \frac{\text{Dsupport}(X \cup Y)}{\text{Dsupport}(X)} - \frac{\text{Dsupport}(Y)}{|D_{Gf}|}$$

The measure can be seen as an estimation of Pr(Y|X) – Pr(Y), which is the increase in probability of Y caused by the occurrence of X. With fuzzy taxonomic structures, it can be extended for fuzzy association rule mining. Briefly speaking, given a threshold R, a rule of interest will be the rule whose Dsupport is more than R times or less than 1/R times of the expected Dsupport (or whose Dconfidence is more than R times or less than 1/R times of the expected Dconfidence). Consider a rule X⇒Y, where X={$x_1$, $x_2$, …, $x_p$} and Y={$y_1$, $y_2$, …, $y_q$}. X^ and Y^ are called the ancestors of X and Y respectively, if $X^\wedge=\{x^\wedge_1, x^\wedge_2, \ldots, x^\wedge_p\}$ (where $x^\wedge_i$ is an ancestor or itself of $x_i$, $1 \le i \le p$) and $Y^\wedge=\{y^\wedge_1, y^\wedge_2, \ldots, y^\wedge_q\}$, (where $y^\wedge_j$ is an ancestor or itself of $y_j$, $1 \le j \le q$). Then the rules $X^\wedge \Rightarrow Y$, $X^\wedge \Rightarrow Y^\wedge$ and $X \Rightarrow Y^\wedge$ are called the ancestors of the rule $X \Rightarrow Y$. Let $\text{Dsupport}_{E(X^\wedge \Rightarrow Y^\wedge)}(X \Rightarrow Y)$ denote the "expected" value of the Dsupport of $X \Rightarrow Y$ on $X^\wedge \Rightarrow Y^\wedge$ and $\text{Dconfidence}_{E(X^\wedge \Rightarrow Y^\wedge)}(X \Rightarrow Y)$ denote the "expected" value of Dconfidence of $X \Rightarrow Y$ on $X^\wedge \Rightarrow Y^\wedge$, then with fuzzy taxonomic structures, we have

$$\text{Dsupport}_{E(X^\wedge \Rightarrow Y^\wedge)}(X \Rightarrow Y) = \frac{\text{Dsupport}(\{x_1\}) \times \ldots \times \text{Dsupport}(\{x_p\})}{\text{Dsupport}(\{x^\wedge_1\}) \times \ldots \times \text{Dsupport}(\{x^\wedge_p\})} \times$$

$$\frac{\text{Dsupport}(\{y_1\}) \times \ldots \times \text{Dsupport}(\{y_q\})}{\text{Dsupport}(\{y^\wedge_1\}) \times \ldots \times \text{Dsupport}(\{y^\wedge_q\})} \times \text{Dsupport}(X^\wedge \Rightarrow Y^\wedge)$$

and

$$\text{Dconfidence}_{E(X^\wedge \Rightarrow Y^\wedge)}(X \Rightarrow Y) = \frac{\text{Dsupport}(\{y_1\}) \times \ldots \times \text{Dsupport}(\{y_q\})}{\text{Dsupport}(\{y^\wedge_1\}) \times \ldots \times \text{Dsupport}(\{y^\wedge_q\})} \times \text{Dconfidence}(X^\wedge \Rightarrow Y^\wedge)$$

With threshold R, the extended measures may be used to filter out redundant rules.

### 2.1.5 Algorithm and Experiments

Accordingly, these extensions have been incorporated into the extended algorithm, and the problem of mining generalized association rules with fuzzy taxonomies and linguistic hedges consist of the following four steps.

1) Transfer the original database D to extended database $D_{Gf}$ ($D_H$) in accordance with user specified $G^f$ and a pool of hedges H.
2) Find all (fuzzy) itemsets whose Dsupports are no less than minimal support. These itemsets are also called frequent itemsets;
3) Use the frequent itemsets as input to generate the rules whose Dconfidences are no less than minimal confidence.
4) Pruning all the uninteresting rules with R-interestingness measures.

Because of the property that $T(a, b) \le a$ and $T(a, b) \le b$, the efficient pruning strategy that any subset of frequent itemsets is also frequent is maintained in generalized association rules with fuzzy taxonomies and linguistic hedge, and can be incorporated into the extended Apriori algorithm. Major differences between our proposed algorithm and traditional GAR algorithm are that: 1) Since an itemset containing two fuzzy items resulting from the same original item is usually considered meaningless (e.g., an itemset containing Young-Age and Old-Age), this may be integrated in the mining process as an optimization strategy. 2) classical sets of items (itemsets) are replaced by fuzzy ones, and fuzzy set operations such as fuzzy intersection, fuzzy union and $\Sigma count$ are used; 3) rules are filtered by R-interestingness measure.

The experiments of our proposed extended Apriori-based mining algorithm to discover association rules with fuzzy taxonomies and linguistic hedges are carried out to verify the effects of fuzziness on computational complexity. The experiments show that the computational complexity is linear with the number of transactions and polynomial to the number of items, which is similar to the non-fuzzy association rule mining. Synthetic experiments revealed that the time consumption of fuzzy association rule mining is stably a bit higher than that of classical method. This is because the incorporation of fuzzy set causes more CPU time and I/O computation on fuzzy degree computation and the generation of the extended database $D_{Gf}$. However, the proposed algorithm is at the same level with computational complexity of GAR.

## 2.2 Implication-Based Fuzzy Quantitative Association Rules

### 2.2.1 Quantitative Association Rules

Though Boolean association rules are meaningful in real-world applications, there are many other situations where data items concerned are usually categorical or quantitative. Examples of such items are Age, Income, Price, Quantity of Product, and so on. Without loss of generality, only consider quantitative items in this section. Apparently, association rules linking quantitative items are meaningful as well, giving rise to so-called quantitative association rules. Usually, quantitative items are represented in a database as attributes whose values are elements of continuous domains such as Real Number Domain R. Such a database is exemplified as D in Table 5.

**Table 5.** Database D with continuous domains

| D | Age | Income |
|---|---|---|
| 111 | 30 | 8500 |
| 112 | 25 | 12500 |
| 113 | 19 | 45000 |
| 114 | 47 | 1500 |
| 115 | 68 | 5000 |

**Table 6.** Database $D_Q$ transformed from D by partitioning domains

| $D_Q$ | Age (0, 30] | Age (30, 60] | Age (60, 100] | Income (0, 5000] | Income (5000, 15000] | Income (15000, ∞) |
|---|---|---|---|---|---|---|
| 111 | 1 | 0 | 0 | 0 | 1 | 0 |
| 112 | 1 | 0 | 0 | 0 | 1 | 0 |
| 113 | 1 | 0 | 0 | 0 | 0 | 1 |
| 114 | 0 | 1 | 0 | 1 | 0 | 0 |
| 115 | 0 | 0 | 1 | 1 | 0 | 0 |

It is easily seen that the typical Apriori algorithm is incapable of dealing directly with such databases for quantitative association rules. Therefore, Srikant & Agrawal 1996 proposed an approach that is composed of two steps: (1) transforming D into a binary database $D_Q$ by partitioning continuous domains, and (2) applying the Apriori algorithm to $D_Q$. For example, if attribute Age takes values from (0, 100], then one could partition (0, 100] into three intervals such as (0, 30], (30, 60], and (60, 100], resulting in three new attributes, namely, Age(0,30], Age(30,60], and Age(60,100] respectively. Likewise, if one partitions the domain of Income into (0, 5000], (5000, 15000], (15000, ∞), then three new attributes related to Income are Income(0, 5000], Income(5000, 15000], Income(15000, ∞). As a result, $D_Q$ becomes a binary database with six attributes as shown in Table 6.

Differently from Boolean AR that represents semantics "Occurrence of Y is associated with Occurrence of X", quantitative AR represents semantics "Quantity of Y is associated with Quantity of X". Formally, for $I = \{I_1, I_2, \ldots, I_m\}$ and D with t being a tuple of D and $t[I_k]$ belonging to a continuous domain ($1 \le k \le m$), suppose that each $I_k$ is partitioned into $p_k$ intervals ($p_k \ge 1$). Then $D_Q$ is with respect to schema $R(I_Q)$ where $I_Q = \{ I_1^1, \ldots, I_1^{p_1}, \ldots, I_k^1, \ldots, I_k^{p_k}, \ldots, I_m^1, \ldots, I_m^{p_m} \}$. For any tuple t in $D_Q$ and $I_k^{p_i}$ in $I_Q$, if $t[I_k]$ in D belongs to interval $p_i$, we have $t[I_k^{p_i}] = 1$, otherwise $t[I_k^{p_i}] = 0$. Then, degrees of support and confidence can be extended directly as traditional association rules as follows:

$$\text{Dsupport}(X \Rightarrow Y) = \text{Dsupport}(X \cup Y) = \frac{\| X \|}{| D_Q |}$$

$$\text{Dconfidence}(X \Rightarrow Y) = \frac{\text{Dsupport}(X \cup Y)}{\text{Dsupport}(X)} = \frac{\| X \cup Y \|}{\| X \|}$$

Where $X, Y \in I_Q$, $X \cap Y = \varnothing$, and $X \cup Y$ does not contain any two items associated with the same original attribute. For example, in table 6, the itemset Age(0,30)∪Age(30,60) will not be considered, and the rule Age(0,30)⇒ Income(5000, 15000) is at Dsupport 40% and Dconfidence 66.7%.

### 2.2.2 Fuzzy Quantitative Association Rules

The sharp boundary of traditional quantitative association rules remains a problem, which may under-emphasize or over-emphasize the elements near the boundaries of intervals in the mining process, and may therefore lead to an inaccurate representation of semantics. This gives rise to a need for fuzzy logic extensions due to the fact that "sharp boundary" is of a typical fuzziness nature. Then, a number of fuzzy sets, usually labeled by linguistic terms, could be defined upon each domain as values of the attributes of D. That is, the intervals in traditional quantitative databases are replaced by fuzzy sets. And fuzzy quantitative association rules express semantic that Y is B is associated with X is A, where A and B are two fuzzy sets defined on domain of X and Y respectively.

Concretely, for $I = \{I_1, I_2, \ldots, I_m\}$ and D with t being a tuple of D and $t[I_k]$ belonging to a continuous domain ($1 \le k \le m$). For each attribute $I_k$, $q_k$ ($q_k \ge 1$) fuzzy sets can be defined on the domain of $I_k$. Then $D_{Qf}$ can be derived accordingly with respect to schema $R(I_{Qf})$ where $I_{Qf} = \{ I_1^1, \ldots, I_1^{p_1}, \ldots, I_k^1, \ldots, I_k^{p_k}, \ldots, I_m^1, \ldots, I_m^{p_m} \}$, and $I_k^{p_i}$ $(1 \leq p_i \leq p_k)$ is a fuzzy set. For any t in $D_{Qf}$ and $I_k$ in I, $t[I_k^{p_i}]$ is the degree that $t[I_k]$ in D belongs to $p_i$th fuzzy set of $I_k$ in D, i.e., $t[I_k^{p_i}] = \mu_{I_k^{p_i}}(t[I_k]) \in [0,1]$ where $\mu_{I_k^{p_i}}$ is the $p_i$th membership function of the attribute $I_k$. That is, tuple t in $D_{Qf}$ supports $I_k^{p_i}$ with a (partial) degree in [0, 1]. With the above extended database $D_{Qf}$.

In our example, if three fuzzy sets Young, Middle and Old are defined on domain of attribute Age and three fuzzy sets Low, Middle and High are defined on domain of attribute Income, new labels (e.g., Young-Age, Middle-Age and Old-Age in place of Age) will be used to constitute a new database $D_{Qf}$ with partial belongings of original attribute values (in D) to each of the new attributes (in $D_{Qf}$). In this way, Table 7 illustrates an example of $D_{Qf}$ obtained from D of Table 5, given fuzzy sets characterized by membership functions shown in Figure 3.

**Table 7.** Database $D_{Qf}$ with fuzzy items

| $D_{Qf}$ | Young-Age | Middle-Age | Old-Age | Low-Income | Middle-Income | High-Income |
|---|---|---|---|---|---|---|
| 111 | 0.8 | 0.7 | 0.2 | 0.5 | 1 | 0.75 |
| 112 | 0.9 | 0 | 0.1 | 0 | 0 | 1 |
| 113 | 1 | 0 | 0 | 0 | 0 | 1 |
| 114 | 0.4 | 1 | 0.6 | 1 | 0 | 0 |
| 115 | 0 | 0 | 1 | 1 | 0 | 0.17 |

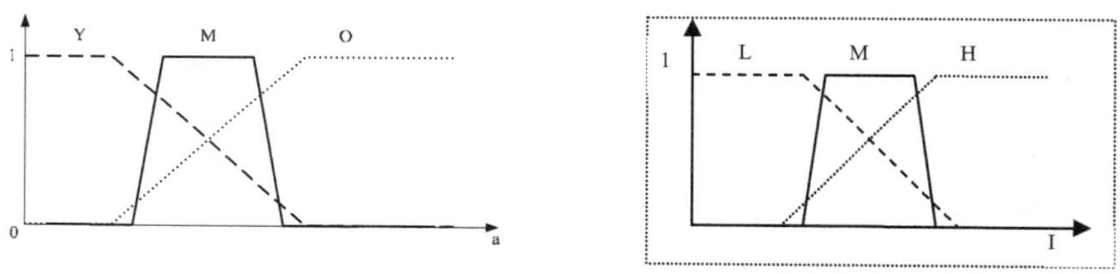


(a) Fuzzy Sets Young(Y), Middle(M) and Old(O) with Y(20, 65), M(25, 32, 53, 60), O(20, 65)

(b) Fuzzy Sets Low(L), Middle(M) and High(H) with L(5000, 12000), M(5000, 6000, 9000, 10000), H(4000, 10000)

**Fig. 3.** Fuzzy sets defined on domain of Age and Income

Then, conventional notions of degrees of support and of confidence can be extended as well. Dsupport of itemset $X=\{x_1, x_2, \ldots, x_p\}$ for a single tuple t can be defined as:

$$\text{Dsupport}_t(X)=T(t(x_1), t(x_2), \ldots, t(x_p))$$

For example, $\text{Dsupport}_{111}(\text{Young-Age} \cup \text{High-Income}) = \min(0.8, 0.75) = 0.75$ if $T(a, b) = \min(a, b)$. Further, Dsupport of itemsets X can be calculated as a generalization of the classical concept of cardinality of a crisp set (De Luca and Termini, 1972):

$$\text{Dsupport}(X) = \frac{\sum count_{t\in D_{Qf}} \text{Dsupport}_t(X)}{|D_{Qf}|}$$

Thus, rule X⇒Y's Dsupport and Dconfidence can be defined as:

$$\text{Dsupport}(X\Rightarrow Y) = \text{Dsupport}(X\cup Y) = \frac{\sum count_{t\in D_{Qf}} \text{Dsupport}_t(X\cup Y)}{|D_{Qf}|}$$

$$\text{Dconfidence}(X\Rightarrow Y) = \frac{\text{Dsupport}(X\cup Y)}{\text{Dsupport}(X)} = \frac{\sum count_{t\in D_{Qf}} \text{Dsupport}_t(X\cup Y)}{\sum count_{t\in D_{Qf}} \text{Dsupport}_t(X)}$$

For the database $D_{Qf}$ in table 7, Dsupport(Young-Age) is equal to (0.8+0.9+1+0.4+0)/5 = 62%. For the selection of T(*a*, *b*) = min(*a*, *b*), rule Young-Age⇒High-Income's Dsupport and Dconfidence can be calculated as Dsupport(Young-Age⇒High-Income) = (min(0.8, 0.75) + min(0.9, 1) + min(1, 1) + min(0.4, 0) + min(0, 0.17))/5 = 2.65/5 = 53%, Dconfidence(Young-Age⇒High-Income) = 2.65/(0.8 + 0.9 + 1 + 0.4 + 0) = 85.5%.

### 2.2.3 Implication-Based Fuzzy Association Rules

As mentioned previously, classically, a rule of X⇒Y is referred to as association between X and Y and modeled by simultaneously appearance and conditional probability for X-to-Y. In classical association rules, Dsupport(X⇒Y) equal to Dsupport(X∪Y) and rules X⇒Y and Y⇒X have the same Dsupport. From a more logic-oriented viewpoint (taking into account the direction of the arrow ⇒), degree of implication (Dimplication) is defined as follows, in which fuzzy implication operators are used (Chen & Wei et al. 1999, Yan & Chen 2005).

$$\text{Dimplication}(X\Rightarrow Y) = \frac{\sum count_{t\in D_{Qf}} I(\text{Dsupport}_t(X),\text{Dsupport}_t(Y))}{|D_{Qf}|}$$

Since I is generally not symmetric, Dimplications of X⇒Y and Y⇒X are generally different. The semantics of rule X⇒Y is the occurrence of X implying the occurrence of Y. For example, with Lukasiewicz operator (I(*a*,*b*) = min(1, 1-*a*+*b*)), Dimplication(Young-Age⇒High-Income) = (min(1, 1-0.8+0.75) + min(1, 1-0.9+1) + min(1, 1-1+1) + min(1, 1-0.4+0) + min(1, 1-0+0.17))/5 = 91%, and Dimplication(High-Income ⇒Young-Age) = (min(1, 1-0.75+0.8) + min(1, 1-1+0.9) + min(1, 1-1+1) + min(1, 1-0+0.4) + min(1, 1-0.17+0))/5 = 94.6%.

Then, an association rule with degrees of support and implication (ARsi), such as X⇒Y, is valid if Dsupport(X⇒Y)≥α, Dimplication(X⇒Y) ≥γ, where α and γ are two given thresholds, X, Y ⊂ $I_{Qf}$, X∩Y = ∅, and X∪Y does not contain any two items associated with the same original attribute. Similar to Dconfidence, Dimplication describes rules strength and can be viewed as a new interestingness measure. For an association rules mining system, users can select the framework of Dsupport-Dconfidence, Dsupport-Dimplication or Dsupport-Dconfidece-Dimplication to represents more flexible semantic information.

Notably, we can prove that for the rule r: X⇒Y, Dsupport(r) ≤ Dconfidence(r) ≤ Dimplication(r). This would be useful for the specification of thresholds. Intuitively, the thresholds should satisfy α≤β≤γ. For example, if α>β, all the rules with Dsupport above $\alpha$ will satisfy the constraint of minimal confidence, which means Dconfidence is useless. Similarly, if $\beta>\gamma$, all the rules with Dimplication more than $\gamma$ and less than $\beta$ will be removed, which means the constraint of Dimplication is useless. Since Dconfidence(r) $\leq$ Dimplication(r), it is also important to note that given $\beta=\gamma$, all the rules discovered in the framework of Dsupport-Dconfidence will also be generated in the framework of Dsupport-Dimplication.

One straightforward method to calculate rules' Dimplication is generating all potential rules from frequent itemsets and scanning the database for each rules' Dimplication. It can be proved that for the proper selection of fuzzy implication operator and t-norm combinations that satisfy $1 + T(a, b) - a = I(a, b)$, which is shown in table 8, the degree of implication can be calculated from degree of support, Dimplication(X⇒Y) = 1 – Dsupport(X) + Dsupport(XY). That is, if itemsets' Dsupport is known, Dimplication can be derived directly from Dsupport, which can be used as a pruning strategy to avoid scanning the database in calculating Dimplication.

**Table 8.** Combinations of t-norms and fuzzy implication operators

| *t-norms* | *Fuzzy implication operators* |
|---|---|
| $T(a, b) = \min(a, b)$ | $I(a, b) = \min(1, 1 - a + b)$ |
| $T(a, b) = ab$ | $I(a, b) = 1 - a + ab$ |
| $T(a, b) = \max(a+b-1, 0)$ | $I(a, b) = \max(1-a, b)$ |

Interestingly, Dubois et al. (2003) also indicated that combinations as shown in the above table build a partition of positive examples, negative examples and irrelevant examples.

Furthermore, recall the notion of simple association rules (SAR) (Chen & Wei et al., 2002), for the above combinations, we have Dimplication(X⇒Y∪Z)= Dimplication(X⇒Y) + Dimplication(X∪Y⇒Z) – 1. This means that Dimplication of the rules with long consequents can be derived from Dimplication of short ones. Let $\Psi$ denote the rule set with Dsupport and Dimplication equal to or greater than given threshold $\alpha$ and $\gamma$ respectively, and $\Psi_s$ denote the rule set in which rules are all of single consequents with $\Psi_s \subseteq \Psi$. It can be proved that $\Psi$ can be derived from $\Psi_s$ according to the above properties and applying given thresholds $\alpha$ and $\gamma$ onto the derived rules will result in exactly the whole rule set. This leads to a substantial reduction of the computational time, fewer rules in the resultant rule set and more interesting rules (Chen & Yan et al., 2004).

Moreover, since $I(\cdot,b)$ is non-increase and $I(a,\cdot)$ is non-decrease, we have Dimplication(X⇒YZ) ≤ Dimplication(X⇒Y) ≤ Dimplication(XZ⇒Y). This is also important for rules' interestingness. For example, if rules X⇒Y and X⇒YZ are all valid association rules (Dsupport, Dimplication and Dconfidence above than $\alpha$, $\beta$ and $\gamma$ respectively), X⇒Y is considered more interesting than X⇒YZ because X⇒Y has larger Dsupport, Dimplication, Dconfidence and is simpler.

According to above discussions, the algorithm of mining implication-based fuzzy quantitative association rules consists of three steps: first transfering the quantitative database to a database $D_{Qf}$ with values on [0, 1] in accordance with fuzzy sets associated with original attributes; second generating all frequent itemsets with extended Apriori algorithm, in which join operation and traditional pruning strategy can also be applied; third, calculating rules' Dimplication (or Dconfidence, or both) and filtering those non-interesting rules, in which above properties can be incorporated.

Experiments on synthetic databases as well as real databases were carried out to show the computational efficiency of the proposed algorithm. The results revealed a remarkable advantage of the proposed algorithm over the straightforward algorithm in computational time. The gap between the two algorithms increases when the number of transactions increases, the number of attributes increases, or $\alpha$ decreases. And the gap remains stable with the change of $\gamma$.

# 3 Mining Functional Dependencies with Degrees of Satisfaction ($FD_d$)

## 3.1 Functional Dependencies with Degrees of Satisfaction ($FD_d$)

Although many attempts have been devoted to discovering the traditional functional dependencies from databases, traditional functional dependencies are generally incapable of dealing with the noise data existent widely in real world applications because classical FD may be too restrictive to hold, since the correspondence of equal X-Y values must be 100% satisfied, by definition. However, it may be meaningful to take into account partial satisfaction of FD, being capable of tolerating the noisy or incomplete/imprecise information at certain degrees.

Recently, we presented the notion of functional dependency with degree of satisfaction (Wei & Chen et al., 2002, Wei & Chen, 2004), which expresses the semantic of equal Y-values corresponding to equal X-values with a certain degree. For a classical database D in schema R(I) where $I = (I_1, I_2, \ldots, I_m)$, and X, Y are collections of attributes (items) in I, then Y is called to functionally depend on X for a tuple pair (t, t') of D, denoted as $_{(t,t')}(X \rightarrow Y)$, if t[X] = t'[X] then t[Y] = t'[Y]. Let $TRUTH_{(t,t')}(X \rightarrow Y)$ denote the truth value that $_{(t,t')}(X \rightarrow Y)$ holds. Apparently, $TRUTH_{(t,t')}(X \rightarrow Y) \in \{0, 1\}$, consistent with truth values of classical logic. In other words, (t, t') satisfies $X \rightarrow Y$ if $TRUTH_{(t,t')}(X \rightarrow Y) = 1$, and (t, t') dissatisfies $X \rightarrow Y$ if $TRUTH_{(t,t')}(X \rightarrow Y) = 0$. Consequently, the degree that D satisfies $X \rightarrow Y$, denoted as $\mu_D(X \rightarrow Y)$, is $TRUTH_D(X \rightarrow Y)$:

$$TRUTH_D(X \rightarrow Y) = \frac{\sum_{\substack{\forall t,t' \in D \\ t \neq t'}} TRUTH_{(t,t')}(X \rightarrow Y)}{|P_D|},$$

where $|P_D|$ is the number of pairs of tuples in D. Clearly, $|P_D| = n(n-1)/2$. Usually, a ($FD_d$) $X \rightarrow Y$ with degree $\alpha$ is denoted as $(X \rightarrow Y)_\alpha$. Given a threshold $\theta \in [0, 1]$, a ($FD_d$) $(X \rightarrow Y)_\alpha$ is valid if $\alpha \geq \theta$. It can be easily seen that FD is a special case of $FD_d$ if $\theta = 1$. As

**Table 9.** Example of Partial Satisfied Functional Dependencies

| ID | Fruits | Drinks |
|---|---|---|
| 1 | Apple | Spirit |
| 2 | Apple | Spirit |
| 3 | Apple | Coca cola |
| 4 | Apple | N/A |
| 5 | Orange | Coca cola |
| 6 | Orange | Coca cola |

an example shown in table 9, a ($FD_d$) Fruits→Drinks can be calculated as $TRUTH_D$(Fruits→Drinks) = [(1+0+0+1+1)+ (0+0+1+1) +(0+1+1) + (1+1) + (1)] / C(6, 2) = 66.7%.

## 3.2 Properties and Mining Methods

One direct way to discover ($FD_d$) is to find all the valid ($FD_d$), which is time consuming. Instead, as we proposed, it can be proven that the following properties hold and some of them can be incorporated into the extended mining algorithm as computational optimization strategies. $\forall$ X, Y, Z $\subseteq$ I. **P1**: If Y $\subseteq$ X, then $TRUTH_D$(X→Y) = 1.

**P2**: $TRUTH_D$(XZ→YZ) ≥ $TRUTH_D$(X→Y).
**P3**: $TRUTH_D$ (X→Z) ≥ $TRUTH_D$ (X→Y) + $TRUTH_D$ (Y→Z) - 1.
**P4**: $TRUTH_D$ (X→Y) + $TRUTH_D$ (Y→Z) ≥ 1.

The task of discovering functional dependencies with degrees of satisfaction could be regarded as discovering all valid $FD_d$ given a threshold θ, θ ∈ [0, 1]. Similar to mining association rules, the mining algorithm is constructed on the lattice as exemplified and shown in figure 4, which could be searched efficiently with a breadth-first strategy.

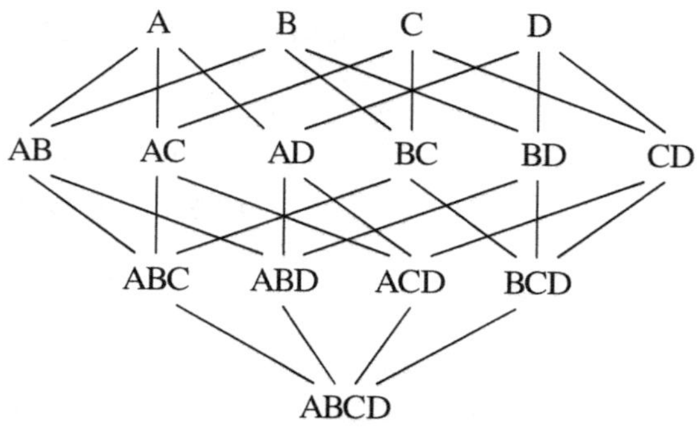


**Fig. 4.** Lattice Structure of Attributes

Let an i-antecedent $FD_d$ be a $FD_d$ with i attributes in the antecedent and a j-consequent $FD_d$ be a $FD_d$ with j attributes in the consequent (i, j = 1, 2, …, m). Then, given the set of all i-antecedent 1-consequent $FD_d$, after filtered with θ, the set of qualified i-antecedent 1-consequent $FD_d$ ($QF_{i1}$) could be derived, based on which the set of candidate i-antecedent 2-consequent $FD_d$ ($CF_{i2}$) could be generated using the property of $TRUTH_D(A \rightarrow B) \geq TRUTH_D(A \rightarrow BX)$. Then $QF_{i2}$ could be filtered out based on $CF_{i2}$. And further $CF_{i3}$ could be generated, and so on until the set of generated candidate i-antecedent $FD_d$ is empty then stop. Thus all qualified i-antecedent $(FDs)_d$ are derived.

In addition, some $FD_d$ in qualified $FD_d$ could be regarded as redundant functional dependencies because of P3. Fox example, given θ = 0.6, $(AB \rightarrow C)_{0.7}$, and $(C \rightarrow D)_{0.9}$, then it could be inferred that $TRUTH_T(AB \rightarrow D) \geq 0.7 + 0.9 - 1 = 0.6$ (P3). Then $(AB \rightarrow D)_\alpha$ could be inferred as a qualified $FD_d$ without scanning the database, so it is can be viewed as redundant knowledge. Furthermore, a minimal set of qualified $FD_d$ can be generated, in which redundant $FD_d$ are filtered (Wei & Chen, 2004).

## 4 Concluding Remarks

Fuzzy association mining has been regarded as a promising area for both researchers and practitioners, due to its advantage in expressing natural language and coping with uncertainty of knowledge. Association rules, functional dependencies, and their fuzzy extensions have been discussed in this chapter. Primary attention has been paid to an overview of our efforts on fuzzy association rules with fuzzy taxonomies, on linguistically modified fuzzy association rules, and on fuzzy implication-based quantitative association rules, as well as on partially satisfied functional dependencies for handling data closeness and noise tolerance.

## References

Agrawal, R., Imielinski, T., Swarmi, A.: Mining Association Rules between Sets of Items in Large Databases. In: Proceedings of the ACM-SIGMOD 1993 International Conference on Management of Data, Washington D.C., pp. 207–216 (1993)

Agrawal, R., Mannila, H., Srikant, R., Toivonen, H., Verkamo, A.I.: Fast Discovery of Association Rules in Advances in Knowledge Discovery and Data Mining. AAAI Press/The MIT Press (1996)

Agrawal, R., Shafer, J.C.: Parallel Mining of Association Rules: Design, Implementation and Experience. Computer Science/Mathematics, RJ 10004 (1996)

Agrawal, R., Srikant, S.: Fast Algorithms for Mining Association Rules. In: Proceedings of the 20th Conference on VLDB, Santiago, Chile (1994); Expanded version available as IBM Research Report RJ 9839

Au, W.H., Chan, K.C.C.: FARM: A Data Mining System for Discovering Fuzzy Association Rules. In: Proc. of the 6th International Conference on Information and Knowledge Management, Las Vegas, Nevada, pp. 209–215 (1997)

Au, W.H., Chan, K.C.C.: An Effective Algorithm for Discovering Fuzzy Rules in Relational Databases. In: Proceedings of the 7th IEEE International Conference on Fuzzy Systems, pp. 1314–1319 (1998)

Aumann, Y., Lindell, Y.: A statistical theory for quantitative association rules. In: Knowledge Discovery and Data Mining, pp. 261–270 (1999)

Bosc, P., Pivert, O.: On some Fuzzy Extensions of Association Rules. In: Proceeding of IFSA/NAFIPS2001 Congress, Vancouver, Canada, pp. 1104–1109 (2001)

Cai, C.H., Fu, A.W., Cheng, C.H., Kwong, W.W.: Mining association rules with weighted items. In: Proceedings of 1998 Intl. Database Engineering and Applications Symposium (IDEAS 1998), Cardiff, Wales, UK, pp. 68–77 (1998)

Castellanos, M., Saltor, F.: Extraction of Data Dependencies. In: Information Modeling and Knowledge bases V (1993)

Chen, G.Q.: Fuzzy Logic in Data Modeling: semantics, constraints and database design. Kluwer Academic Publishers, Boston (1998)

Chen, G.Q., Ai, J., Yu, W.: Discovering Temporal Association Rules for Time-Lag Data. In: Proceedings of International Conference on e-Business (ICEB 2002), p. 324 (2002)

Chen, G.Q., Wei, Q., Kerre, E.E.: Fuzzy Data Mining: Discovery of Fuzzy Generalized Association Rules. In: Recent Research Issues on Management of Fuzziness in Databases, in the Physica-Verlag series. Studies in Fuzziness and Soft Computing. Springer, Heidelberg (1999)

Chen, G.Q., Wei, Q.: Fuzzy Association Rules and the Extended Mining Algorithms. Information Sciences 147, 201–228 (2002)

Chen, G.Q., Wei, Q., Liu, D., Wets, G.: Simple Association Rules (SAR) and the SAR-Based Rule Discovery. Computer & Industrial Engineering 43(2002), 721–733 (2002)

Chen, G.Q., Wei, Q., Zhang, H.: Discovering Similar Time-Series Patterns with Fuzzy Clustering and DTW Methods. In: IFSA/NAFIPS 2001, Vancouver (2001)

Codd, E.F.: A Relational Model for Large Shared Data Banks. Communications of the ACM 13(6), 377–387 (1970)

Chen, G.Q., Yan, P., Kerre, E.E.: Computationally Efficient Mining for Fuzzy Implication-Based Association Rules in Quantitative Databases. International Journal of General Systems 33, 163–182 (2004)

Chien, B.C., Lin, Z.L., Hong, T.P.: An Efficient Clustering Algorithm for Mining Fuzzy Quantitative Association Rules. In: Proceedings of the 9th International Fuzzy Systems Association World Congress, Vancouver, Canada, July 25-28, pp. 1306–1311 (2001)

Cubero, J.C., et al.: Data Summarization in Relational Databases through Fuzzy Dependencies. Information Sciences 121(3-4), 233–270 (1999)

Cubero, J.C., Medina, J.M., Pons, O., Vila, M.A.: Rules discovery in fuzzy relational databases. In: Conference of the North American Fuzzy Information Processing Society, NAFIPS 1995, Maryland, USA, pp. 414–419. IEEE Computer Society Press, Los Alamitos (1995)

De Cock, M., Cornelis, C., Kerre, E.E.: Fuzzy association rules: A two sided approach. In: Proceedings of international conference on fuzzy information processing theories and applications (2003)

De Luca, A., Termini, S.: A definition of Non-Probabilistic Entropy in the Setting of Fuzzy Sets. Inform. Control 20, 301–312 (1972)

Dubois, D., Hullermeier, E., Prade, H.: A note on Quality Measures for Fuzzy Association Rules. In: Proceeding of IFSA 2003 (2003)

Fayyad, U., Uthurusamy, R.: Efficient Algorithms for Discovering Association Rules. In: AAAI Workshop on Knowledge Discovery in Databases, Seattle, Washington, DC, USA, July 1994, pp. 181–192 (1994)

Fayyad, U., Piatetsky-Shapiro, G., Smyth, P.: From Data Mining to Knowledge Discovery: An Overview. In: Fayyad, U., Piatetsky-Shapiro, G., Smyth, P., Uthurusamy, R. (eds.) Advances in Knowledge Discovery and Data Mining, pp. 1–30. AAAI Press/The MIT Press, Cambridge (1996)

Fu, A., et al.: Finding fuzzy sets for the mining of fuzzy association rules for numerical attributes. In: Proceedings of 1st Intl. Symposium on Intelligent Data Engineering and Learning (IDEAL 1998), pp. 263–268 (1998)

Fukuda, T., Morimoto, Y., Morishita, S.: Data Mining Using Two-Dimensional Optimized Association Rules: Scheme, Algorithms, and Visualization. In: Proc. of the 1996 ACM-SIGMOD Int'l Conf. on the Management of Data, pp. 12–13 (1996)

Graff, J.M., Kosters, W.A., Witteman, J.J.W.: Interesting Fuzzy Association Rules in Quantitative Databases. In: Siebes, A., De Raedt, L. (eds.) PKDD 2001. LNCS, vol. 2168, pp. 140–151. Springer, Heidelberg (2001)

Gyenesei, A.: A fuzzy approach for mining quantitative association rules, TUCS technical reports 336, University of Turku, Department of Computer Science, Lemminkisenkatu14, Finland (2000a)

Gyenesei, A.: Mining Weighted Association Rules for Fuzzy Quantitative Items. In: Proceedings of PKDD Conference, pp. 416–423 (2000b)

Gyenesei, A., Teuhola, J.: Interestingness Measures for Fuzzy Association Rules. In: Siebes, A., De Raedt, L. (eds.) PKDD 2001. LNCS, vol. 2168, pp. 152–164. Springer, Heidelberg (2001)

Han, J., Fu, Y.: Discovery of Multiple-level Association Rules from Large Databases. In: Proceedings of the 21st International Conference on Very Large Databases, Zurich, Switzerland (1995)

Han, J., Kamber, M.: Data Mining: Concepts and Techniques. Morgan Kaufmann Publishers, San Francisco (2000)

Hong, T.P., Kuo, C.S., Chi, S.C.: A fuzzy data mining algorithm for quantitative values. In: The Third International Conference on Knowledge-Based Intelligent Information Engineering Systems, pp. 480–483 (1999a)

Hong, T.P., Kuo, C.S., Chi, S.C.: Mining association rules from quantitative data. Intelligent Data Analysis 3(5), 363–376 (1999b)

Houtsma, M., Swarmi, A.: Set Oriented Mining of Association Rules, Technical Report RJ 9567, IBM (1993)

Huhtala, Y., Karkkainen, J., Paokka, P., Toivonen, H.: TANE: An Efficient Algorithm for Discovering Functional and Approximate Dependencies (1998a), `http://citeseer.nj.nec.com/`

Huhtala, Y., Karkkainen, J., Porkka, P., Toivonen, H.: Efficient Discovery of Functional and Approximate Dependencies Using Partitions. In: Proc. 14th Int. Conf. on Data Engineering. IEEE Computer Society Press, Los Alamitos (1998b)

Hullermeier, E.: Implication-Based Fuzzy Association Rules. In: ECML/PKDD 2001, Freiburg, Germany (2001a)

Hullermeier, E.: Fuzzy Association Rules: Semantics Issues and Quality Measures (2001b), `http://citeseer.nj.nec.com/`

Klemettinen, M., Mannila, H., Ronkainen, P., Toivonen, H., Verkamo, A.I.: Finding Interesting Rules from Large Sets of Discovered Association Rules. In: Proceedings of Third International Conference on Information and Knowledge Management (1994)

Klir, G.J., Yuan, B.: Fuzzy sets and fuzzy logic theories and applications. Prentice Hall PTR, Englewood Cliffs (1995)

Kruse, R., Nanck, D., Borgelt, C.: Data Mining with Fuzzy Methods: Status and Perspectives (2001), `http://citeseer.nj.nec.com/`

Kuok, C.M., Fu, A., Wong, M.H.: Mining Fuzzy Association Rules in Databases. SIGMOD Record 27(1), 41–46 (1998)

Liao, S.Y., Wang, H.Q., Liu, W.Y.: Functional Dependencies with Null Values, Fuzzy Values, and Crisp Values. IEEE Transactions on Fuzzy Systems 7(1), 97–103 (1999)

Luo, J.X.: Integrating Fuzzy Logic with Data Mining Methods for Intrusion Detection, Thesis of Master Degree of Science in the Department of Computer Science, Mississippi State University (1999)

Luo, J.X., Bridges, S.M.: Mining fuzzy association rules and fuzzy frequency episodes for intrusion detection. International Journal of Intelligent Systems (IJIS) 15(8), 687–703 (2000)
Maimon, O., Kandel, A., Last, M.: Information-Theoretic Fuzzy Approach to Knowledge-Discovery in Databases (2001), http://citeseer.nj.nec.com/
Mannila, H., Toivonen, H., Verkamo, A.I.: Efficient Algorithms for Discovering Association Rules. In: AAAI Workshop on Knowledge Discovery in Databases, Seattle, Washington, pp. 181–192 (1994)
Mazlack, L.J.: Approximate Clustering in Association Rules. In: 19th International Conference of the North American Fuzzy Information Processing Society - NAFIPS 2000, Atlanta, pp. 256–260 (2000)
Mueller, A.: Fast sequential and parallel algorithms for association rule mining: A comparison. Technical Report CS-TR-3515, Dept. of Computer Science, Univ. of Maryland, College Park, MD (1995)
Rifqi, M., Monties, S.: Fuzzy Prototypes for Fuzzy Data Mining (2001), http://citeseer.nj.nec.com/44393.html
Savasere, E., Omiecinski, S., Navathe: An Efficient Algorithm for Mining Association Rules in Large Databases. In: Proceedings of the VLDB Conference, Zurich, Switzerland (1995)
Savnik, I., Flach, P.A.: Discovery of Multi-valued Dependencies from Relations, report00135 (2000), http://citeseer.nj.nec.com/savnik00discovery.html
Shu, J., Tsang, E., Yeung, D.S., Shi, D.: Mining fuzzy association rules with weighted items. In: Proc. IEEE Int'l Conf. on System, Man and Cybernetics (SMC 2000), Nashville, Tennessee (2000)
Srikant, R., Agrawal, R.: Mining Generalized Association Rules. In: Proc. of the 21st Int'l Conference on Very Large Databases, Zurich, Switzerland (1995)
Srikant, R., Agrawal, R.: Mining Quantitative Association Rules in Large Relational Tables. In: SIGMOD 1996 6/96 Montreal, Canada (1996)
Srikant, R., Vu, Q., Agrawal, R.: Mining Association Rules with Item Constraints. In: Proc. of the 3rd Int'l Conference on Knowledge Discovery in Databases and Data Mining, Newport Beach, California, USA (1997)
Tseng, S.M.: Mining Association Rules with Interestingness Constraints in Large Databases (2001), http://citeseer.nj.nec.com/
Wei, Q., Chen, G.Q.: Mining Generalized Association Rules with Fuzzy Taxonomic Structures. In: 18th Int'l Conf. of NAFIPS, New York, NY, USA, pp. 477–481 (1999)
Wei, Q., Chen, G.Q., Kerre, E.E.: Mining Functional Dependencies with Degrees of Satisfaction in Databases. In: Proceedings of Joint Conference on Information Sciences, Durham, NC, USA (2002)
Wei, Q., Chen, G.Q.: Efficient Discovery of Functional Dependencies with Degrees of Satisfaction FDd. International Journal of Intelligent Systems 19, 1089–1110 (2004)
Yan, P., Chen, G.Q.: Discovering a Cover Set of ARsi with Hierarchy from Quantitative Databases. Information Sciences 173(4), 319–336 (2005)
Yang, Y.P., Singhal, M.: Fuzzy Functional Dependencies and Fuzzy Association Rules (2001), http://citeseer.nj.nec.com/
Yilmaz, E., Triantaphyllou, E., Chen, J., Liao, T.W.: A heuristic for mining association rules in polynomial time. In: Mathematical and Computer Modeling (2002)
Zadeh, L.: Fuzzy Sets. Information and Control 8, 338–358 (1965)
Zhang, L., Chen, G.Q., Brijs, T., Zhang, X.: Discovering During-Temporal Patterns (DTPs) in Large Temporal Databases. Expert Systems with Applications 34, 1178–1189 (2008)

# 5

# Dempster-Shafer Structures, Monotonic Set Measures and Decision Making

Ronald R. Yager

Machine Intelligence Institute, Iona College
New Rochelle, NY 10801
yager@panix.com

**Abstract.** We first formulate the problem of decision making under uncertainty. The importance of the representation of our knowledge about the uncertainty in formulating a decision process is pointed out. We provide a brief discussion of the case of probabilistic uncertainty. Next, in considerable detail, we discuss the case of decision making under ignorance. For this case the fundamental role of the attitude of the decision maker is noted and its subjective nature is emphasized. Next the case in which a Dempster-Shafer belief structure is used to model our knowledge of the uncertainty is considered. Here we also emphasize the subjective choices the decision maker must make in formulating a decision function. The case in which the uncertainty is represented by a monotonic set measure is then investigated. We then return to the Dempster-Shafer belief structure and show its relationship to the set measure. This relationship allows us to get a deeper understanding of the formulation the decision function used Dempster- Shafer framework. We discuss how this deeper understanding allows a decision analyst to better make the subjective choices needed in the formulation of the decision function. Finally we provide a generalized framework for decision-making in the face of Dempster-Shafer type uncertainty.

**Keywords:** Decision Making, Uncertainty, Dempster-Shafer Belief Structure, Monotonic Set Measures. OWA Operators, Choquet Integral.

## 1 Introduction

While historically probability has been the predominate method for modeling the knowledge about the value of an uncertain variable modern technology has recently provided us with a rich selection of additional formalisms for representing this kind of knowledge [1]. Among these are possibility theory, Dempster-Shafer theory of evidence and monotonic set measures. In a large part the interest in these additional formalisms has been motivated by the interest in building computationally intelligent systems which benefit from the inclusion of various types of human sourced knowledge. Three types of uncertainty can easily be seen to appear in human sourced knowledge, randomness, granularity and gradualarity [2]. Rather then being competitive these various representations are useful for modeling these different types of uncertainties and situations regarding our knowledge of the uncertainty.

In many applications where our objective is the selection of a best course of action from a set of available alternatives there exists some uncertainty regarding the value

E. Rakus-Andersson et al. (Eds.): Recent Advances in Decision Making, SCI 222, pp. 67–88.
springerlink.com 

of a variable which effects the results obtained from a selection of a course of action. This task is referred to as **D**ecision **M**aking **U**nder **U**ncertainty (DMUU). Here our uncertainty formalisms are used to model the knowledge about these relevant variables. An often-used approach for comparing the alternatives is to use a valuation function. Using a valuation function we generate for each alternative a single value, called the representative value, and then compare the alternatives with respect to these representative values. The determination of the representative value of an alternative depends upon the payoffs associated with the alternative, our knowledge about the uncertain variables and the decision attitude of the decision maker. The construction of the valuation function is strongly dependent upon the representation used for modeling our knowledge about the uncertain variable. In the case of probabilistic uncertainty a well-known example of representative value is the expected value. Other examples of representative value in the probabilistic are the median and the mode. Often the choice between these different representative values is a subjective one depending on the attitude of the decision-making. An important determining feature is the decision maker's attitude with respect to being optimistic or pessimistic (aggressive or conservative).

Our focus here, for the most part, will be on the Dempster-Shafer belief structure [3] and particularly the problem of decision making in the face of this type of uncertainty. This framework is particularly useful when modeling uncertain knowledge that has both randomness and granularity. As we shall see the Dempster-Shafer framework has very rich connections with the probabilistic and set measure frameworks for representing uncertainty. One view of the D-S belief structure is as extension probability theory in which the probabilities rather then precisely known are only known to lie within intervals. Here we clearly see the confluence of randomness and granularity. Another interpretation of the D-S structure is related to the set measure formalism for modeling uncertainty. Here it can be used as a formalism for modeling partial knowledge about what is the appropriate set measure in a given situation.

Here we shall look at the issue of constructing valuation functions for the case of decision making with D-S uncertainty and particularly take advantage its connections with probability theory as well as set measure models.

## 2 Decision Making under Uncertainty

In figure 1, the $A_i$ are a collection of possible actions open to a decision maker. The $x_j$ are a set of possible values for the state of some relevant variable U. $C_{ij}$ is the payoff to the decision maker if he selects alternative $A_i$ and the state of U is $x_j$. Our decision task is the selection of the alternative that provides the decision maker with the best payoff. Often this choice must be made in situations where the decision maker does not have complete knowledge of the value of U, it is DMUU.

In DMUU rather than knowing the unique payoff resulting from the selection of an alternative we know the collection of possible payoffs that can be obtained if we select an alternative. In this situation the problem of comparing the alternatives becomes very difficult, as we must compare multi-dimensional objects. One way to compare the alternatives is to associate with each alternative a single value, $V(A_i)$, called its representative value. The alternatives can then be compared using these representative values. In order to avoid "gamesmanship" in formulation of available alternatives, including an alternative simply because it help another alternative, we shall require that the calculation of the representative value satisfy Arrow's requirement of indifference to irrelevant alternatives [4]. This indifference can be guaranteed if the representative value of each alternative is calculated without using any data about other alternatives. Because of this independence in the calculation of each alternative's valuation we can focus on a generic alternative A with payoff $C_j$ resulting from the case when $U = x_j$.

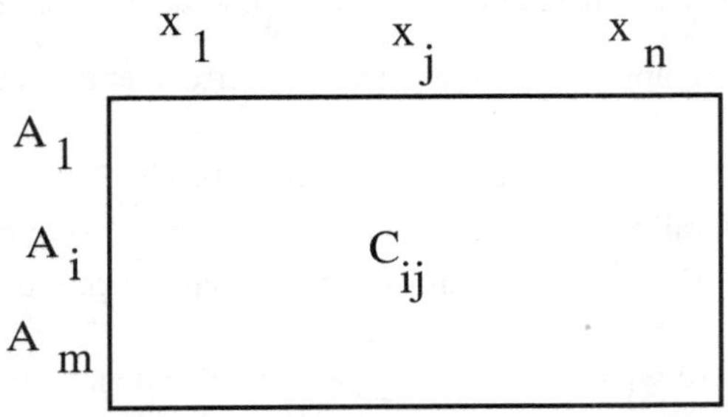


**Fig. 1.** Decision Making Paradigm

Two situations regarding our knowledge about the uncertainty of payoffs have been well studied in the literature [5]. One of these is the case in which we have no information about the underlying mechanism determining the value from X, **D**ecision **M**aking **U**nder **I**gnorance (DMUI). The other situation is one in which the underlying mechanism determining the value of U is assumed to be probabilistic. In this case $p_j$ is the probability that $x_j$ will be the value of U. In the case of probabilistic uncertainty the most common approach is to use the expected value as the representative valuation, $V(A) = E(A) = \sum_{j=1}^{n} C_j \, p_j$.

While the expected value (mean) is the most common approach to evaluating an alternative in the face of probabilistic uncertainty other methods exist. One of these is to use the mode. We recall the mode takes as its value the payoff with the highest probability. More formally if p-index is a mapping such that p-index(j) is the $j^{th}$ largest of the probabilistic than Mode(A) = $C_{p\text{-index}(1)}$, it is the payoff in the highest probability.

Another method used in the probabilistic setting to obtain the valuation of an alternative is the median. The median makes use of the cumulative distribution function, CDF. We recall that

$$CDF(z) \equiv \sum_{\substack{j \text{ s.t} \\ C_j \leq z}} p_j$$

It is essentially the probability that the payoff will be less than or equal z. The median is defined as the payoff for which there is a 50% chance of getting equal or above.

## 3 OWA Operators and Decisions under Ignorance

In the case of decision making under ignorance (DMUI) the evaluation of the representative value is based on the decision attitude of the responsible decision maker [5]. An optimistic decision maker evaluates alternative A as $V(A) = Max_j[C_j]$. A pessimistic decision maker evaluates $E(A) = Min_j[C_j]$. A neutral decision maker uses $V(A) = \frac{1}{n}\sum_{j=1}^{n} C_j$. Another approach, suggested by Arrow and Hurwicz [5], calculates $V(A) = \alpha\, Max_j[C_j] + (1 - \alpha)\, Min_j[C_j]$ where $\alpha \in [0, 1]$.

In [6] Yager generalized these approaches using the OWA operator [7, 8]. We recall that $OWA(a_1, ..., a_n) = \sum_j w_j b_j$ where $b_j$ is the $j^{th}$ largest of the $a_i$ and $w_j$ are a collection of weights where $w_j \in [0, 1]$ and $\sum_{j=1}^{n} w_j = 1$. We denote the n vector W whose components are the $w_j$ as the OWA weighting vector. We shall say that a vector having the properties $w_j \in [0, 1]$ and $\sum_{j=1}^{n} w_j = 1$ is a proper vector.

In [6] it was suggested calculating the valuation as $V(A) = OWA(C_1, ..., C_n)$. By appropriately selecting W we can obtain the preceding valuations. For $W = W^*$ where $w_1 = 1$ and $w_j = 0$ for all $j \neq 1$ we get the optimistic valuation. For $W = W_*$ where $w_n = 1$ and $w_j = 0$ for $j \neq n$ we get the pessimistic valuation. For $w_j = 1/n$ we get the neutral. For W such that $w_1 = \alpha$, $w_n = 1 - \alpha$ and $w_j = 0$ for all others we get the Arrow-Hurwicz formulation.

In addition to providing the already established valuation procedures this generalization allowed for the consideration of other methods. For example one can consider the "olympic average." Here we eliminate the highest and lowest values and take the average of the rest, $w_1 = 0$, $w_n = 0$ and $w_j = \frac{1}{n-2}$ for the others. A generalized form of this olympic average is to eliminate the q highest and lowest. In this case: $w_j = 0$ for $j = 1$ to $q$ & $j = n - q + 1$ to $n$ and $w_j = \frac{1}{n-2q}$ for all the rest. A median type valuation method can be modeled using this formulation, if n is odd then we take $w_{\frac{n+1}{2}} = 1$ and $w_j = 0$ for all others and if n is even then $w_{\frac{n}{2}} = w_{\frac{n}{2}+1} = 0.5$ and $w_j = 0$ for all others.

An alternative expression of the OWA aggregation OWA($C_1$, ..., $C_n$) will be useful. If (j) is the index of the $j^{th}$ largest of the payoffs then we can express OWA($C_1$, ..., $C_n$) $=\sum_j w_j C_{\pi(j)}$.

The $w_j$ can be viewed (interpreted) as the "probability" that the $j^{th}$ best outcome will occur. We note that for an optimist $w_j = 1$, he believes the probability that the *best* thing will happen is one. A pessimist has $w_n = 1$, he believes that the probability that the *worst* thing will happen is one. Thus we see the decision attitude of the decision is captured by the $w_j$. We shall refer to these as "attitudinal probabilities." Under this interpretation of the weights V(A) can be viewed as a kind of expected value.

Two characterizing measures can be associated with the attitudinal vector W. [7]. The first, called the **attitudinal character**, is defined as A-C(W) = $\frac{1}{n-1}\sum_{j=1}^{n}(n-j)\, w_j$. It can be easily shown that A–C(W) $\in$ [0, 1]. Further we note that A-C($W^*$) = 1 and A-C($W_*$) = 0 and when W has $w_j = \frac{1}{n}$ then A-C(W) = 0.5. Within the framework of using the OWA operator to evaluate an uncertain alternative the value of A-C(W) can be interpreted as providing a measure of the degree of optimism associated with valuation process.

A second measure introduced in [7] is Disp(W) = $-\sum_{j=1}^{n} w_j \ln[w_j]$. Under the interpretation of $w_j$ as attitudinal probabilities Disp(W) can be seen as the entropy of W.

In the OWA approach the form of the valuation function determined by weights . In the following we describe some methods for obtaining the weights that allow for the consideration of cognitively expressed preferences of the decision maker and are expressed in a manner independent of the cardinality of vector.

A very popular method is due to O'Hagan [9]. In this method the decision maker need only provide a degree of optimism $\alpha$. Using this degree of optimism we solve the following mathematical programming problem for the $w_j$.

$$\textbf{Max: } -\sum_{j=1}^{n} w_j \ln(w_j)$$

**s|t**

1. $\frac{1}{n-1}\sum_{j=1}^{n}(n-j)w_j = \alpha$

2. $\sum_{j=1}^{n} w_j = 1$

3. $0 \le w_j \le 1$

We call this the Maximum Entropy method and refer to the weights obtained using this method as the ME-OWA weights for a given $\alpha$.

Another approach to the determination of the OWA weights was introduced by Yager [10]. This makes use of a class of functions which we called BUM, functions. A BUM function is a mapping $f:[0, 1] \rightarrow [0, 1]$ such that $f(0) = 0$, $f(1) = 1$ and $f(x) \geq f(y)$ if $x > y$. Using these functions we obtain the OWA weights as $w_j = f(\frac{j}{n}) - f(\frac{j-1}{n})$ for $j = 1$ to $n$.

A fundamental motivation behind the use of these BUM functions is that they can allow us to model different cognitive preferences of a decision maker. For example we can use parameterized classes of these functions in which the parameters are chosen so as to enforce some predilection of the decision maker. Another possible use of these BUM functions is within the spirit of Zadeh's idea of computing with words [11]. Here with the aid of fuzzy set theory we can use these functions to represent some linguistically expressed specifications.

We note that the area of under f(x), $\int_0^1 f(x)dx$, provides a useful approximation to the degree of optimism of any weighting vector generated from the BUM function f(x), thus if W is the vector generated from f then $\int_0^1 f(x)dx \approx \alpha(W)$. Thus by specifying the area under f(x) we are essentially characterizing the attitude of the resulting valuation function.

We can also provide a relationship between the weight generating function f and a kind of attitudinal cumulative distribution function, A-CDF. Let F be a CDF associated with the decision maker's perception of the payoff to be received. Specifically for any $r \in [0, 1]$, F(r) indicates the decision maker's belief that the probability that at least r percentage of the available payoffs are less than or equal to the actual payoff received. Thus if F(0.5) = 0.8 then the decision makers believes that there is 0.8 probability that half of the available payoffs will be less then the actual payoff received. In figure #2, we provide such an A-CDF. Here the decision-making is indicating that mainly the higher available payoffs will be bigger than the actual payoff received.

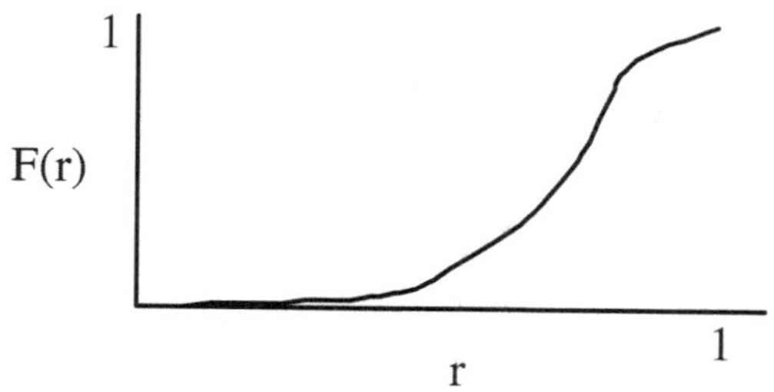


**Fig. 2.** Cumulative distribution function

Using F we can obtain our weight generating function as f(x) = 1 - F(1 - x). Since the weights are generated from f we can obtain the weights directly for F as $w_j = F(\frac{n+1-j}{n}) - F(\frac{n-j}{n})$.

Thus here the relevant decision makers can provide this A-CDF and then generate the weights.

## 4 Dempster-Shafer Belief Structures

Formally a Dempster-Shafer belief structure [3, 12] is a set mapping m: $2^X \rightarrow [0, 1]$ such that $m(\varnothing) = 0$ and $\sum_{B \in 2^X} m(B) = 1$. The function m is called the basic assignment function and the collection of subsets $B_j$ such that $m(B_j) \neq 0$ are called the focal elements. A number of different semantics can be associated with this structure. The one that we shall find useful here is the one related to the random set point of view. Here we have a variable U that can take a value in the space X. We are uncertain about the actual value of the variable. Our knowledge about the value of the variable U can be modeled in terms of the following random experiment. We perform an experiment whose outcome is a subset of the space X where $m(B_j)$ is the probability that the outcome is the subset $B_j$. We emphasize that the ***outcome*** is a subset of X. Once having obtained a subset B* as a result of this experiment an element is selected from B* as the value of U. However, the method of selection of the element from B* is unknown.

Two important set measures have been associated with a D-S belief structure. For any subset A of X, the measure of plausibility is defined as $Pl(A) = \sum_{B_j \cap A \neq \varnothing} m(B_j)$ and the measure of belief is defined as $Bel(A) = \sum_{B_j \subseteq A} m(B_j)$. It can be shown that $Pl(A) \geq Bel(A)$.

It is easily illustrated that the D-S framework provides a representation of a situation in which our knowledge about the probability of the elements in X are intervals rather than specific values. Under this interpretation of the D-S belief structure the measure Pl(A) is the upper probability of the subset A and the measure Bel(A) is the lower probability of the subset A. Thus the probability of the subset A, Prob(A), is bounded as follows $Pl(A) \leq Prob(A) \leq BelA)$. Thus one use of the D-S belief structure is to provide a generalization of probability theory where our knowledge of the probabilities of events are not precisely known but only know within intervals.

We now return to our concern with decision-making and consider the situation in which our knowledge about the uncertain variable U is expressed in terms of a Dempster-Shafer belief structure. Here we describe an approach to decision making suggested by Yager [6].

Let the knowledge of U be expressed by a D-S belief structure m with q focal elements, $B_j$ for j = 1 to q. We shall let $n_j$ denote the cardinality of $B_j$. Viewing the D-S structure as a situation in which we choose one of the $B_j$ with probability $m(B_j)$ we can consider the following framework for valuating alternatives. For alternative A we let $V_j(A)$ denote the valuation of A in the case in which we obtained focal element $B_j$. Using this notation the overall valuation of alternative A is

$$V(A) = \sum_{j=1}^{q} V_j(A)\, m(B_j),$$

it is the expected value of the valuations of A under each of the focal elements.

The next issue is the determination of $V_j(A)$, the valuation of alternative A in the situation in which we obtained $B_j$. As we indicated in the D-S framework the method of selection of the element from $B_j$ is unknown. Thus we see that here we are faced with a situation of decision making under ignorance. In order to obtain the valuation $V_j(A)$ we must use one of the methods from decision making under ignorance. Specifically if $A(B_j)$ are the set of payoffs associated with outcomes in $B_j$ then we can evaluate $V_j(A) = OWA(A_j(B_j))$. The use of the OWA operators requires the selection of some weight vector to reflect the decision maker's attitude. We shall let $W_j$ indicate the vector associated with $B_j$, this is a vector of dimension $n_j$ whose components, $w_j(i)$, lie in the unit and sum to one.

In [6] we suggested a unified approach to the determination of the vector $W_j$. We suggested that the decision maker supply a degree of optimism $\alpha$ and then obtain the weighting vector $W_j$ for each focal element by solving the ME-OWA mathematical programming problem for the weights. This approach requires the decision maker to provide only one parameter. In the following example we illustrate this approach.

**Example:** Consider a decision problem in which we have an uncertain variable U which can take a value in the space $X = \{x_1, x_2, x_3, x_4, x_5\}$. Assume that our knowledge about the value of this uncertain variable is represented by a D-S belief structure m with three focal elements:

$$B_1 = \{x_1, x_3, x_4\},\ B_2 = \{x_2, x_5\} \text{ and } B_3 = \{x_1, x_2, x_3, x_4, x_5\}$$

where $m(B_1) = 0.6$, $m(B_2) = 0.3$ and $m(B_3) = 0.1$. Assume the payoff matrix associated with alternative **A** is shown below:

| | $x_1$ | $x_2$ | $x_3$ | $x_4$ | $x_5$ |
|---|---|---|---|---|---|
| **A** | 7 | 5 | 12 | 13 | 6 |

Using focal elements and the payoff matrix we obtain the collection of payoffs associated with each focal element: $A(B_1) = <7, 12, 13>$, $A(B_2) = <5, 6>$ and $A(B_3) = <7, 5, 12, 13, 6>$.

Finally we shall assume that our decision has expressed a degree of optimism of 0.75. Solving the appropriate ME-OWA mathematical program problem we obtain the weights associated with the OWA operator for an optimism value of 0.75 under various argument cardinalities.

| # of arguments | w(1) | w(2) | w(3) | w(4) | w(5) |
|---|---|---|---|---|---|
| **2** | .0.75 | .0.25 | | | |
| **3** | 0.62 | 0.27 | 0.11 | | |
| **4** | 0.52 | 0.27 | 0.14 | 0.07 | |
| **5** | 0.46 | 0.26 | 0.15 | 0.08 | 0.05 |

Using this can calculate the valuation of $B_j$, $V_j(A) = OWA(A(B_j)$.

$V_1(A) = OWA(7, 12, 13)$ using the vector $W^T = [0.6, 0.27, 0.11]$ we get

$V_1(A) = (.62)(13) + (.27)(12) + (.11)(7) = 12.07$

$V_2(A) = OWA(5, 6)$ using vector $W^T = [0.75, 0.25]$ we get

$V_2(A) = (.75)6 + (.25)5 = 5.75$

Finally $V_3(A) = OWA(7, 5, 12, 13, 6)$ using vector $W^T = [0.46, 0.26, 0.15, 0.08, 0.05]$ we get

$$V_3(A) = (13)(.40) + (12)(2.6) + (7)(.15) + (6)(.08) + (5)(.05) = 10.88$$

Combining these we get

$$V(A) = \sum_j V_j(A)m(B_j) = (0.6)(12.07) + (0.3)(5.75) + (0.1)(10.88) = 10.055$$

We note that this approach combines the technique of probabilistic decision making with the technique decision making under ignorance. The calculation of $V_j(A)$ is obtained using the technique of decision making under ignorance, $V_j(A) = OWA(A(B_j))$. In order to perform this calculation we needed a $|B_j|$ dimension OWA weighting vector $W_j$. In the approach developed in [6] we suggested obtaining $W_j$ using ME-OWA approach with a specified degree of optimism $\alpha$. The weighting vectors need not be obtained this way. At a formal level all we need is any collection of q proper weighing vectors each of the appropriate dimension. From a pragmatic point of view the vector should reflect the decision maker's attitude. Thus in this approach we need a collection $W_1$, ....., $W_q$ of proper vectors which we shall denote as the attitudinal vectors. We note that any of the other techniques introduced earlier for obtaining the weights can be used in place of the ME–OWA approach.

## 5 Monotonic Set Measures

We now consider another framework for representing our knowledge about an uncertain variable. This approach makes use of the idea of a monotonic set measure, sometimes called a fuzzy measure [3, 13].

**Definition:** A mapping $\mu: 2^X \rightarrow [0, 1]$ is called a monotonic set measure on the space X if it has the properties: **1.** $\mu(\emptyset) = 0$, **2.** $\mu(X) = 1$ and **3.** $\mu(A) \leq \mu(B)$ if $A \subset B$

If U is an uncertain variable a set measure $\mu$ can be used to represent our knowledge about the value of the variable by using $\mu(A)$ to indicate "our confidence" that the value of U lies in the set A. In some settings more specific names are given to $\mu$.

Various different types of well-established uncertainty frameworks can be expressed using this formalism. Probability theory is a special case where our measure is additive, $\mu(A \cup B) = \mu(A) + \mu(B)$ if $A \cap B = \emptyset$. Here $\mu$ corresponds to a probability measure.

One useful generic view of $\mu(A)$, which will help our intuition, is as a generalization of the concept of Prob(A), only here we don't necessarily require additivity.

Possibilistic uncertainty [14, 15] is another special case of monotonic measures, here we have $\mu(A \cup B) = \text{Max}[\mu(A), \mu(B)]$.

The situation of complete certainty, where we know $V = x^*$, is represented by a measure where $\mu(A) = 1$ if $x^* \in A$ and $\mu(A) = 0$ if $x^* \notin A$. Actually this can be seen to be a special case of both a probability and possibility measure.

Another special class of measures are the cardinality-based measures where $\mu(A)$ just depends on the number of elements in A. In particular for these measures no information is available distinguishing the elements. Three important special cases of this cardinality type measure are $\mu^*$, $\mu_*$ and $\mu_N$.

For $\mu^*$: $\mu^*(\emptyset) = 0$ and $\mu^*(A) = 1$ for all $A \neq \emptyset$.

For $\mu_*$: $\mu_*(X) = 1$ and $\mu_*(A) = 0$ for all $A \neq X$

For $\mu_N$: $\mu_N(A) = \frac{|A|}{n}$, where n is the cardinality of X.

We note that our use of set measures is not limited to these well-known situations but can be used to model other situations. It provides a very general framework for modeling knowledge about uncertain variables.

The use of a set measure to capture our knowledge about a variable requires that we have the value of $\mu(A)$ for **all** subsets A. This requires $2^n$ pieces of data. This is a very strong requirement and is often very difficult to satisfy. An important feature of the three set measures, probability, possibility and cardinality is that their special structures allows a great simplification in the amount of information required. In each of these special cases we only need no more then n pieces of data.

For probability we need the n probabilities, $p_i = \mu(\{x_i\})$, and since $\sum_i^n p_i = 1$ we need only specify $n - 1$ of these. For possibility measures we need the n possibilities $\alpha_i = \mu(\{x_i\})$ and require that $\text{Max}_i[\alpha_i] = 1$. For the cardinality based measures we need $h_i$, the measure of a set of cardinality i. We note that $h_i \geq h_j$ if $i > j$ and that $h_0 = 0$ and $h_n = 1$.

The technique for evaluating an alternative when the information about the underlying variable is expressed by a monotonic set measure makes use of the Choquet integral [16, 17]. Here again we shall consider a generic alternative A with payoff $C_i$ resulting when the variable $U = x_i$. In this case we shall assume that our knowledge about the variable U is expressed by a set measure $\mu$ on the space $X = \{x_1, ..., x_n\}$ of possible outcomes. In the following we let *index* be a function such that index(j) is the index of the $j^{th}$ largest payoff. Hence $C_{index(j)}$ is the $j^{th}$ largest payoff and $x_{index(j)}$ is the corresponding value of the output variable. Let $H_j = \{x_{index(i)}$ for $i = 1$ to $j\}$, it is the set of outcomes associated with the $j^{th}$ largest payoffs. Using this we obtain the valuation of alternative A as

$$V(A) = \sum_{j=1}^{n} w_j \, C_{index(j)}$$

where $w_j = \mu(H_j) - \mu(H_{j-1})$. It should be noted that $w_j \in [0, 1]$ and $\sum_{j=1}^{n} w_j = 1$.

In order to help get some feel for the appropriateness of this we shall show that this is a generalization of the expected value used in probability theory. Consider the situation of a probabilistic distribution with payoffs $C_1, ..., C_n$ and associated probabilities $p_i$. Here the expected value is $E(A) = \sum_{i=1}^{n} p_i \, C_i$ . Again letting index(j) be the index of the j largest payoff we can rewrite this as $E(A) = \sum_{j=1}^{n} p_{index(j)} \, C_{index(j)}$.

Letting $H_j = \{x_{index(j)}/\text{for } i = 1 \text{ to } j\}$ then $Prob(H_j) = \sum_{i=1}^{j} p_{index(i)}$. Using this we can express $P_{index(j)} = Prob(H_j) - Prob(H_{j-1})$. Here we see that $E(A) = \sum_{i=1}^{n} p_i \, C_i = \sum_{j=1}^{n} w_j C_{index(j)}$ where $w_j = Prob(H_j) - Prob(H_{j-1})$.

Thus we see that the Choquet formulation generalizes the expected value where we replace $Prob(H_j)$ with $\mu(H_j)$. In the special case when $\mu$ is a probability measure the two approaches are the same. Here again we see the correspondence between the probability of a subset and the measure of the subset.

We observe the situation for the case of a cardinality based measures where $\mu(A) = h_{|A|}$. Thus here we have a collection $h_1 > ... > h_n$ of weights such that $h_0 = 0$ and $h_n = 1$. In this case since $H_j = \{x_{index(j)}\ j = 1 \text{ to } j\}$ has j elements we see

$w_j = \mu(H_j) - \mu(H_{j-1}) = h_j - h_{j-1}$. Denoting $\Delta_j = h_j - h_{j-1}$ we get $V(A) = \sum_{j=1}^{n} \Delta_j C_{index(j)}$. We note that the $\Delta_j$ are collection of weights such that $\Delta_j \in [0, 1]$ and $\sum_{j=1}^{n} \Delta_j = 1$.

## 6 Alternative Set Measure Valuation Methods

With our appreciation of the correspondence between monotonic set measures and the probability measure we can begin to look at some ideas used in probability theory and see their corresponding manifestation in the more general framework of monotonic set measures.

In probability theory one alternative to the use the expected value as a method of valuation is to use the median. Let us obtain a median type valuation operator in the case in which our knowledge of an uncertain value is described by a monotonic set measure.

In probability theory the cumulative distribution function, CDF, is defined such that CDF(y) is the probability that the payoff will be at least y. We further recall that for a random variable the median is the payoff value where the CDF either equals or transitions past the value 0.5. We now generalize this idea to the case of set measures. We again let $H_j = \{x_{index(j)}$ for i = 1 to j$\}$, the set of outcomes with the j highest payoffs. For the subset $H_j$, the measure $\mu(H_j)$ can be viewed as the confidence (generalized probability) that we shall receive a payoff at least as big as the $j^{th}$ payoff. We now see the correspondence between the CDF and the measures on the family of $H_j$ functions. Using this correspondence we can extend the idea of median to the set measure. We let k be the index such that $\mu(H_k) \geq 0.5$ and $\mu(H_{k-1}) < 0.5$. Then we define the median valuation of the payoffs as $C_{index(k)}$.

Another measure used in probability theory is the mode. In probability the mode is the outcome most likely to occur, it is the value with the highest probability. Again let $H_j = \{x_{index(j)}$ for j = 1 to j$\}$ and $w_j = \mu(H_j) - \mu(H_{j-1})$. We define the mode as $C_{index(r)}$ such that $w_r = \text{Max}_j[w_j]$. We observe that if $\mu$ is a probability measure, where $w_j = p_j$, then the mode is as desired the payoff with the largest probability.

We see that if $\mu$ is the special measure $\mu^*$, then since $w_1 = 1$ and $w_j = 0$ for all $j \neq 1$ the mode of $\mu^*$ is $C_{index(1)}$, it is the biggest payoff. When $\mu$ is $\mu_*$ then since $w_n = 1$ and $w_j = 0$ for all $j \neq n$ the mode is $C_{index(n)}$, it is the smallest payoff.

More generally when $\mu$ is a cardinality-based measure the mode is the payoff value associated with the position having the largest weight.

Another important measure associated with a probability distribution is variance. Here we shall extend this idea to the situation in which we have a monotonic set measure. Again assume a space $X = \{x_1, ..., x_n\}$ with a measure $\mu$ defined on this space. For A such that $C_i$ is the value associated with the outcome $x_i$ we defined its expected value as $E_\mu(A) = \sum_{k=1}^{n} w_j C_{index(j)}$ with index(j) the index of j largest payoff and $w_j = \mu(H_j) - \mu(H_{j-1})$ where $H_j = \{x_{index(i)} | i = 1 \text{ to } j\}$.

We now define the variance. Here we let $d_k = (C_k - E_\mu(A))^2$. Furthermore we define d–index as a function of the $d_k$ such that d-index(j) is the index of the $j^{th}$ largest of $d_k$. Before defining the variance we introduce

$$G_i = \{x_{d\text{-}index(k)} | k = 1 \text{ to } i\},$$

it is the set of i outcomes with the largest value for $d_k$. Furthermore we let $g_i = \mu(G_i) - \mu(G_{i-1})$. Using this we define variance of A as

$$Var_\mu(A) = \sum_{i=1}^{n} g_i \, d_{d\text{-}index(i)}$$

It is the Choquet integral of variable $(C_k - E_\mu(A)^2$ with respect to the measure $\mu$.

Let us consider some special cases. First is the case where $\mu$ is a probability measure. In this case $E_\mu(A)$ is the expected value, $\overline{C}$. Thus $d_k = (C_k - \overline{C})^2$. Here

$$g_i = \mu(G_i) - \mu(G_{i-1}) = p_{d\text{-}index(i)},$$

it is the probability of the outcome with the ith largest deviation from the mean. Thus

$$Var_\mu(A) = \sum_{i=1}^{n} p_{d\text{-}index(i)} d_{d\text{-}index(i)} = \sum_{j=1}^{n} P_j (C_j - \overline{C})^2 = E(C^2) - (\overline{C})^2$$

Here then we get as desired the usual variance.

Consider now the case when $\mu$ is $\mu^*$, $\mu(B) = 1$ for $B \neq \varnothing$ and $\mu(B) = 0$ for $B = \varnothing$. In this case $\mu(H_1) = 1$ and hence $w_1 = 1$ and $w_j = 0$ for $j \neq 1$. Thus $E_\mu(A) = C_{index(1)} = Max_i[C_i] = C^*$, it is the largest payoff. Consider now the calculation of $Var_\mu(A)$. Here $d_k = (C_k - C^*)^2$. Since $\mu = \mu^*$ then independent of the components $\mu(G_j) = 1$ for $j > 0$ and $\mu(\varnothing) = 0$. Hence $g_1 = 1$ and $g_j = 0$ for all $j > 1$. Thus $Var_\mu(A) = \sum_{j=1}^{n} g_j \, d_{d\text{-}index(j} = d_{d\text{-}index(1)}$. It is equal to the largest $(C_k - C^*)^2$ this occurs when $C_k$ is the $Min_j[C_j] = C_{index(n)}$ Thus

$$Var_\mu(A) = d_{d\text{-}index(1)} = (C_{index(n)} - C^*)^2 = (C_* - C^*)^2$$

where $Min_j[C_j] = C_*$ and $Max_j[C_j] = C^*$.

Consider now the case when $\mu = \mu_*$, here $\mu(B) = 0$ for $B \neq X$ and $\mu(X) = 1$. Here then $\mu(H_j) = 0$ for $j \neq n$ and $\mu(H_n) = 1$ thus $w_j = 0$ for $j \neq n$ and $w_n = 1$ hence

$$E_\mu(\overline{A}) = \sum_{j=1}^{n} w_j C_{index(j)} = C_{index(n)} = Min_i[C_i] = C_*$$

Consider now $Var_\mu(A) = \sum_{j=1}^{n} g_j\, d_{d\text{-}index(j}$. Here since $\mu(G_j) = 0$ for $j \neq n$ and $\mu(G_n) = 1$ we get $g_j = 0$ for $j \neq n$ and $g_n = 1$ thus $Var_\mu(A) = d_{d\text{-}index(n)}$, the smallest of the d values. This occurs when $d_{d\text{-}index(n)} = (C_* - C_*)^2$. Here $Var_\mu(A) = 0$. Thus in this case we have no variance.

We note that we can express

$$Var_\mu(A) = \sum^{n} g_i(C_{d\text{-}index(i)} - \overline{C})^2 = \sum^{n} g_i C^2_{d\text{-}index(i)} - 2\overline{C}\sum^{n} g_i C_{d\text{-}index(i)} + \overline{C}^2$$

where d-index(i) is ordered in decreasing order by the distance of $C_{d\text{-}index(i)}$ from $\overline{C}$. In the case where the measure is a cardinality based measure, $g_i = w_i$, we get

$$Var_\mu(A) = \sum_{i=1}^{n} w_i\, C^2_{d\text{-}index(i)} - 2\overline{C}\sum_{i=1}^{n} w_i\, C_{d\text{-}index(i)} + \overline{C}^2$$

where $\overline{C} = \sum_{i=1}^{n} w_i C_{d\text{-}index(i)}$. Our two examples are easily seen as special cases of this.

## 7 Monotonic Set Measures and D-S Structures

While the monotonic set measure provides a very general framework for the representation of knowledge about an uncertain variable, it can impose a heavy burden with respect to the required data. Often we are only able to provide part of the information needed. This can lead to a situation in which, rather than uniquely specifying the measure associated with a variable, we can only identify a subset of measures among which the appropriate one lies.

In this section we return to the Dempster-Shafer belief structure and look at its relationship to monotonic set measures. As we shall see the D-S belief structure can be viewed as providing a framework for indicating an imprecise specification of the measure associated with a variable. In particular a D-S belief structure on X can be seen to be associated with a subset of possible monotonic set measures on X.

Let m be a belief structure on X with focal elements $B_j$, $j = 1$ to q. It is well known that the plausibility measure, $Pl(A) = \sum_{B_j \cap A \neq \Phi} m_j(B_j)$, and belief measure, $Bel(A) = \sum_{B_j \subseteq A} m(B_j)$ are monotonic set measures. It is also established that $Pl(A) \geq$

Bel(A). In [18] Yager described a fundamental relationship between the Dempster-Shafer belief structure and set measures. He suggested that a D-S belief structure could be used to provide a representation of our knowledge about a variable when there exists some uncertainty regarding our knowledge of the underlying set measure.

More specifically assume U is a variable on X having some specific, but unknown, associated monotonic measure $\mu$. Let **M** be the set of all possible monotonic set measures on the space X. Without any further information we have no knowledge about the measure associated with U other than $\mu \in$ **M**.

As discussed in [18] the knowledge conveyed by a D-S belief constrains the set of possible measures that can be associated with U. It reduces the set **M** to a subset **S**. Thus a D-S belief structure can be viewed as providing ***partial*** information about the actual measure associated with the variable U.

A procedure was described in [18] for generating the subset **S** of monotonic set measures associated with a belief structure m. We now describe this procedure. Assume m is a D-S belief structure with q focal elements. Let $B_j$ be a focal element of m with cardinality $|B_j| = n_j$. We first introduce the idea of an **allocation vector**. A valid allocation vector for $B_j$ is an $|B_j|$ dimensional vector $W_j$ with components $w_j(i)$ having the properties: **1.** $w_j(i) \in [0, 1]$ and **2.** $\sum_{i=1}^{|B_j|} w_j(i) = 1$. The key feature distinguishing the allocation vectors associated with the different $B_j$ is the cardinalities of the vectors.

An **allocation bundle** is a q-tuple $\mathbf{W} = \langle W_1, ..., W_q \rangle$ where $W_j$ is a valid allocation vector for $B_j$. Thus an allocation bundle is a collection of allocation vectors one for each of the focal elements.

If m is a D-S belief structure and **W** an allocation bundle then the set function $\mu$ defined by

$$\mu(E) = \sum_{j=1}^{q} (m(B_j) \sum_{i=1}^{|B_j \cap E|} w_j(i))$$

is a monotonic set measure: $\mu(\emptyset) = 0$, $\mu(X) = 1$ and $\mu(E) \geq \mu(F)$ if $E \subset F$. Thus each allocation bundle corresponds to a monotonic measure in the space of possible measures for the variable.

A few notable examples of these allocation bundles are worth pointing out. The first is $\mathbf{W}^*$. For this bundle all the $W_j$ are such that $w_j(1) = 1$. The first element in each allocation vector is 1. It can be shown that this allocation bundle corresponds to the plausibility measure, Pl. Another special case is $\mathbf{W}_*$ here the $w_j$ are such that $w_j(n_j) = 1$ for all j. The last element in each allocation vector is 1. This allocation bundle is the belief measure. In the following we shall find it convenient to denote the monotone measure generator from the allocation bundle **W** as $\mu_{\mathbf{W}}$. If we need to also emphasize the D-S structure we shall use $\mu_{\mathbf{W}/m}$, where m is the D-S structure. Here

$\mu_{\mathbf{W}}(A)$ is the measure of the subset A for the measure generated by **W**. It can be shown [18] that for any allocation bundle **W** its associated set measure $\mu_{\mathbf{W}}$ satisfies

$$\mu_{\mathbf{W}^*} = Pl(A) \geq \mu_{\mathbf{W}}(A) \geq Bel(A) = \mu_{\mathbf{W}_*}$$

Thus we see that **W*** and $\mathbf{W}_*$ induce the bounding set measures associated with a given belief structure m. This observation is in support of the central role that the belief and plausibility play in the D–S theory. These measures provide the extremes in the space of potential measures.

Another interesting and important allocation bundle is the one in which each $W_j$ is such that $w_j(i) = 1/n_j$. For this measure

$$\mu(E) = \sum_{j=1}^{q} m(B_j)\frac{|B_j \cap E|}{|B_j|} = \sum_{j=1}^{q} \frac{m(b_j)}{|B_j|} \text{Min}[n_j, |E|]$$

This is the measure used by Smets in his work [19]. For this measure

$$\mu(\{x_i\}) = \sum_{j=1}^{n} \frac{1}{n_j} m(B_j)B_j(x_i) = \sum_{j \text{ s.t } x_i \in B_j} \frac{1}{n_j}(m(B_j))$$

It can be shown this is a probabilistic measure, it is additive.

While there is no requirements on the relationship between the individual allocation vectors making up an allocation bundle, bundles in which there exists some consistency or relationship between the individual vector would seen to be more interesting. Consistent classes of allocation bundles can be obtained using a BUM function f: $[0, 1] \rightarrow [0, 1]$. We recall such a function has $f(0) = 0$, $f(1) = 1$ and $f(x) \geq f(y)$ if $x \geq y$. We shall denote $\mathbf{W}_f$ as allocation bundle generated from f. For these types of bundles we define the individual allocation vector $W_j$ such that

$$w_j(i) = f\left(\frac{i}{n_j}\right) - f\left(\frac{i-1}{n_j}\right).$$

## 8 A View of Decision Making with D-S Structures

Using the results of the preceding section we can provide a deeper understanding of the process we earlier introduced for decision making where the uncertainty was expressed using a D-S belief structure. This understanding will further help us in the process of choosing a particular set measure.

Let us recall the situation in which we have an alternative with payoff $C_i$ associated with outcome $x_i$ . We have a D-S belief structure m on the outcome space X which has focal elements $B_j$ for j = 1 to q.. In our approach we associated with each $B_j$ an OWA weighting vector $\widetilde{W}_j$ called the attitudinal vector. This is a vector of dimension of dimension $|B_j|$, $n_j$, its components $\widetilde{w}_j(i)$ sum to one and lie in the unit interest. Here the weighting vector reflected our decision attitude. Thus $\widetilde{W}_j$ is a proper vector of dimension $n_j$.

In our approach we calculated $V_j(A) = OWA(A(B_j))$ and then obtained

$$V(A) = \sum_{j=1}^{q} m(B_j)V_j(A).$$

Here we recall $A(B_j)$ is the collection of payoffs associated with the outcomes in focal set $B_j$.

Let *j-index* be an index defined on the set $B_j$. Specifically j-index(k) is the index of the kth largest payoff in the set $A(B_j)$. Thus $C_{j\text{-index}(k)}$ is the $k^{th}$ largest payoff in $A(B_j)$. Using this notation $V_j(A) = \sum_{k=1}^{n_j} \widetilde{w}_j(k)C_{j\text{-index}(k)}$ and hence $V(A) = \sum_{j=1}^{q}\left(m(B_j)\sum_{k=1}^{n_j} \widetilde{w}_j(k)C_{j\text{-index}(k)}\right)$.

Let us now consider another approach to making decisions with D-S belief structures. As we have earlier indicated we can use a D-S structure to model our knowledge in the situation in which we only have partial knowledge of the set measure associated with the variable. The use of a D-S belief structure corresponds to associating with the variable a subset of possible set measures. Since, with the aid of the Choquet integral, we know how to evaluate a decision alternative where our knowledge about the outcome variable is represented by a set measure one approach to evaluating an alternative in the face of D-S uncertainty is to select a representative set measure from the family of possible measures and use this to generate the representative value of the alternative. If $\mu$ is the selected measure then

$$E(A) = \sum_{k=1}^{n} (\mu(H_k) - \mu(H_{k-1}))C_{index(k)} = \sum_{k=1}^{n} h_k C_{index(k)}$$

In this situation, where our knowledge is a D-S belief structure, any $\mu$ selected must be one that can be generated from the belief structure m using some allocation Bundle. Thus the selection of $\mu$ effectively corresponds to the selection of an allocation bundle. Let $\mathbf{W} = \langle W_1, ..., W_q \rangle$ be the selected bundle. In this case $\mu(E) = \sum_{j=1}^{q} (m(B_j)) \sum_{i=1}^{|B_j \cap E|} w_j(i))$.

Consider now $\mu(H_k)$ where $H_k$ is the set of outcome with the k highest values payoffs

$$\mu(H_k) = \sum_{j=1}^{q} (m(B_j)) \sum_{i=1}^{|B_j \cap E_k|} w_j(i))$$

We further see that

$$h_k = \mu(H_k) - \mu(H_{k-1}) = \sum_{j=1}^{q} (m(B_j) \sum_{i=1}^{|B_j \cap H_k|} w_j(i) - \sum_{=1}^{q} (m(B_j) \sum_{i=1}^{|B_j \cap H_{k-1}|} w_j(i)$$

Thus

$$h_k = \sum_{j=1}^{q} (m(B_j) \left[ \sum_{i=1}^{|B_j \cap H_k|} w_j(i) - \sum_{j=1}^{|B_1 \cap H_{K-1}|} w_j(i) \right] =- \sum_{j=1}^{q} m(B_j)\, g_j(k)$$

Let us carefully look at $g_j(k)$.

In the following we shall let $n_{j/k} = |B_j \cap H_k|$. It is the number of the k highest payoffs in $B_j$ or the number of elements of $B_j$ among the k highest payoffs under A. First we note that either $n_{j/k} = n_{j/k-1}$ or $n_{j|k} = n_{j|k-1} + 1$. If $n_{j/k} = n_{j/k-1}$ then $g_j(k) = 0$ and if $n_{j|k} = n_{j|k-1} + 1$ then $g_j(k) = w_j(n_{j|k})$

As we noted that $n_{j|k}$ is the number of the k highest payoffs in $B_j$ or the number of elements of $B_j$ among the k highest payoffs. Clearly $g_j(k) = 0$ if $k \geq n_j$. Furthermore we note that if outcome corresponding to $C_{index(k)}$ is not in $B_j$ then $g_j(k) = 0$. Specifically $g_j(k) \neq 0$ if $C_{index(k)}$ is in $B_j$. Furthermore $g_j(k) = w_j(i)$ if $C_{index(k)}$ is the $i^{th}$ largest element in the subset $B_j$. Let us recall that $C_{j\text{-}index(i)}$ is the $i^{th}$ largest element in $B_j$.

Let us now return to our formulation for E(A). We see from the above that

$$E(A) = \sum_{k=1}^{n} h_k C_{index(k)} = \sum_{k=1}^{n} \left( \sum_{j=1}^{n} m(B_j) g_j(k) C_{index(k)} \right)$$

We can rearrange to terms so that $E(A) = \sum_{j=1}^{n} m(B_j) \left( \sum_{k=1}^{n} g_j(k) C_{index(k)} \right)$.

However as we have noted $g_j(k) = 0$ if $x_{index(k)} \notin B_j$ and $g_i(k) = w_j(i)$ if $C_{index(k)} = C_{j\text{-}index(i)}$. Using this we get

$$E(A) = \sum_{j=1}^{n} m(B_j) \left( \sum_{i=1}^{n_j} w_j(i) C_{j\text{-}index(i)} \right)$$

From this result we see that if $w_j(i) = w_j(i)$ then E(A) =V(A). The implication of this is that essentially both approaches are the same. That is the method used in [6] essentially corresponds to selection of a measure. Thus if $W_1, ..., W_q$ are a collection of proper vectors of dimension $n_j$ respectively than using this as an allocation bundle to choose a particular measure and then using the Choquet integral to evaluate this measure is the same as using this collection of vectors to provide the attitudinal OWA vectors and directly evaluate the belief structure. With this understanding the collection **W** = $<W_1, ..., W_q>$ can be seen more generally as providing some resolution of the uncertainty with respect to the measure that is appropriate. Thus we shall refer to **W** $<W_1, ..., W_q>$, as an A–bundle, instead of using the terms allocation or attitudinal

## 9 Generalized Decision Making with D-S Structures

We previously noted that the mode and median provide alternatives to the expected value and provided their formulations in the case of a set measure. We look at the use of these in the situation in which our information about the outcome variable is provided in terms of a D-S belief structure, we have partial information about the set measure. We note a fundamental distinction should be made between the mode and median on one hand and the expected value on the other. The mode and median always provide a value that is one of the payoff values. The mean, on the other hand, returns as its value some blending of the payoffs. Thus the mode and median are what we call **celibate**, they don't combine payoffs.

Assume we have some belief structure m on X with q focal elements $B_j$. Let **W** = $\{W_1, ..., W_q\}$ be an A-bundle, each $W_j$ is a proper matrix of dimension $n_j = |B_j|$. Let $C_i$ be the payoffs associated with the alternative being evaluated, $C_i$ being associated with outcome $x_i$. Let index(k) be the index of the $k^{th}$ largest payoff.

We now associate with each focal element an n dimensional vector $R_i$, n is the dimension of the set X. We denote the $k^{th}$ element in $R_j$ as $R_j(k)$. For each $B_j$ we construct $R_j$ as follows:

**1.** Initialize k = 1 and i = 1
**2.** If $x_{index(k)} \notin B_j$ then set $R_j(k) = 0$
**3.** If $x_{index(k)} \in B_j$ set Open(i) = k
set i = i + 1
**4.** if k = n go to .5
if k < n, set k = k + 1 go to 2
**5.** For i = 1 to $n_j$
$R_j(open(i)) = w_j(i)$

Essentially $R_j$ is a n vector such that its kth component corresponds to index(k). It has zero in all positions for which $x_{index(k)}$ is not contained in $B_j$. In all positions corresponding to those which there is an element in the focal element it has an appropriate weight. We note that $R_j$ is a proper vector of dimension n. We shall let **R** denote a matrix whose columns are the $R_j$.

We let **M** be the column vector of dimension q such its jth component is $m(B_j)$ Let shall denote the product **E** = **R M**. Thus **E** is an n dimension vector, we shall denote the $k^{th}$ element as E(k).

As we have already noted the choice of an A-bundle **W** essentially "selects" a set measure, $\mu_{\mathbf{W}}$. We can easily show that

$$E(k) = h_k = \mu_{\mathbf{W}}(H_k) - \mu_{\mathbf{W}}(H_{k-1})$$

Using this we can obtain the mode and median. To obtain the mode, we simply find k* such that $k^* = \underset{k=1 \text{ to } n}{\text{Max}} [E(k)]$ and then let Mode(A) = $C_{index(k^*)}$

Likewise the median can be easily obtained. Let $S_j = \sum_{k=1}^{j} E(k)$. We then obtain r such that $S_r \geq 0.5$ and $S_{r-1} < 0.5$ using this we obtain Median(A) = $C_{index(r)}$.

The introduction of the $R_j$ vector provides a very unifying framework. Using these we can easily express the Choquet valuation. Here we let **C** be the n vector such that its kth component is $C_{index(k)}$, the kth largest payoff under the alternative A being evaluated. Using this we get

$$VAL(A) = \mathbf{C}^T(\mathbf{RM})$$

Actually our original procedure for evaluating belief structures, $\tilde{V}(A) = \sum_{i=1}^{q} V_j(A)m(B_j)$, can be easily expressed using this form and its equivalence to the Choquet method can be easily seen. First we note that $V_j(A) = C^T R_j$ and hence $\tilde{V}(A) = \sum_{i=1}^{q} \mathbf{C}^T R_j m(B_j) = \mathbf{C}^T(\mathbf{RM})$

One *mode like* method of evaluation that makes itself apparent in the framework of the D-S belief structure is the following. Let m-index(j) be the focal element with the $j^{th}$ largest weight. Thus $m(B_{m\text{-}index(j)})$ is the $j^{th}$ largest weight. We can consider a valuation of A as $V_{m-index(1)}(A)$. That is we take as the valuation of alternative A the valuation of the focal element with the largest weight. We denote this as $\widetilde{Mod}_1(A)$. We note that this doesn't necessarily result in a valuation equal to one of the argument payoffs. Thus its spirit is not necessary in the idea of the mode.

There exists an interesting view of this $\widetilde{\text{Mod}}_1(A)$ evaluation that easily relates it to the Choquet valuation method. First we note that we can very naturally express this method using our $R_j$ vectors. As we indicated $V_j(A) = \mathbf{C}^T R_j$. Let us define $\text{Val}_\alpha(A)$ as

$$\text{Val}_\alpha(A) = \sum_{j=1}^{q} V_j(A) \frac{m(B_j)^\alpha}{\sum_{K=1}^{q} m(B_j)^\alpha}$$

Here we see that if $\alpha \to \infty$ then $\text{Val}_\alpha(A) = \widetilde{\text{Mod}}_1(A)$. Furthermore if $\alpha = 1$ then this becomes the original valuation method. If we let $\mathbf{M}^\alpha$ be the q dimensional vector whose j component is

$\dfrac{m(B_j)^\alpha}{\sum_{i=1}^{q} m^\alpha(B_i)}$ then we can express $\text{Val}_\alpha(A)$ as $\mathbf{C}^T\mathbf{R}\mathbf{M}^\alpha$.

## 10 Conclusion

We formulated the problem of decision-making under uncertainty and noted the importance of the structure used for representing the uncertainty in formulating the decision process. We briefly looked at the case of probabilistic uncertainty. Next we investigated the case of decision making under ignorance. In this case the fundamental role of the attitude of the decision maker was noted and its subjective nature was emphasized. We then considered the case in which a Dempster-Shafer belief structure is used to model our knowledge of the uncertainty and emphasized the subjective choices the decision maker must make in formulating a decision function. The case in which the uncertainty is represented by a monotonic set measure was then investigated. We then returned to the Dempster-Shafer belief structure and showed its relationship to the set measure. This relationship allowed us to get a deeper understanding of the formulation the decision function used in the Dempster-Shafer environment.

## References


[1] Klir, G.J.: Uncertainty and Information. John Wiley & Sons, New York (2006)

[2] Zadeh, L.A.: Generalized theory of uncertainty (GTU)-principal concepts and ideas. Computational Statistics and Data Analysis 51, 15–46 (2006)

[3] Yager, R. R., Liu, L. (Dempster, A.P., Shafer, G., (advisory eds.)): Classic Works of the Dempster-Shafer Theory of Belief Functions. Springer, Heidelberg (2008)

[4] Arrow, K.J.: Social Choice and Individual Values. John Wiley & Sons, New York (1951)

[5] Luce, R.D., Raiffa, H.: Games and Decisions: Introduction and Critical Survey. John Wiley & Sons, New York (1967)

[6] Yager, R.R.: Decision making under Dempster-Shafer uncertainties. International Journal of General Systems 20, 233–245 (1992)

[7] Yager, R.R.: On ordered weighted averaging aggregation operators in multi-criteria decision making. IEEE Transactions on Systems, Man and Cybernetics 18, 183–190 (1988)
[8] Yager, R.R., Kacprzyk, J.: The Ordered Weighted Averaging Operators: Theory and Applications. Kluwer, Norwell (1997)
[9] O'Hagan, M.: Using maximum entropy-ordered weighted averaging to construct a fuzzy neuron. In: Proceedings 24th Annual IEEE Asilomar Conf. on Signals, Systems and Computers, Pacific Grove, CA, pp. 618–623 (1990)
[10] Yager, R.R.: Quantifier guided aggregation using OWA operators. International Journal of Intelligent Systems 11, 49–73 (1996)
[11] Zadeh, L.A.: Fuzzy logic = computing with words. IEEE Transactions on Fuzzy Systems 4, 103–111 (1996)
[12] Shafer, G.: A Mathematical Theory of Evidence. Princeton University Press, Princeton (1976)
[13] Murofushi, T., Sugeno, M.: Fuzzy measures and fuzzy integrals. In: Grabisch, M., Murofushi, T., Sugeno, M. (eds.) Fuzzy Measures and Integrals, pp. 3–41. Physica-Verlag, Heidelberg (2000)
[14] Zadeh, L.A.: Fuzzy sets as a basis for a theory of possibility. Fuzzy Sets and Systems 1, 3–28 (1978)
[15] Dubois, D., Prade, H.: Possibility Theory: An Approach to Computerized Processing of Uncertainty. Plenum Press, New York (1988)
[16] Grabisch, M., Murofushi, T., Sugeno, M.: Fuzzy Measures and Integrals. Physica-Verlag, Heidelberg (2000)
[17] Choquet, G.: Theory of Capacities. Annales de l'Institut Fourier 5, 131–295 (1953)
[18] Yager, R.R.: A class of fuzzy measures generated from a Dempster-Shafer Belief Structure. International Journal of Intelligent Systems 14, 1239–1247 (1999)
[19] Smets, P., Kennes, R.: The transferable belief model. Artificial Intelligence 66, 191–234 (1994)

6

# The Development of Interpretable Decision-Making Models: A Study in Information Granularity and Semantically Grounded Logic Operators

Witold Pedrycz

Department of Electrical & Computer Engineering
University of Alberta, Edmonton T6R 2G7 Canada
Systems Research Institute, Polish Academy of Sciences
Warsaw, Poland

**Abstract.** The fundamental feature of human-friendly decision-making models (such as those encountered in complex medical problems, economical or political systems, technical diagnostic of physical systems, etc.) is predominantly concerned with interpretability of resulting constructs. Interpretability comes hand in hand with the granular nature of conceptual entities which are sought as the generic building blocks of such decision models and directly support a logic nature of their processing. From the system development perspective, the interpretability begs for solutions to the fundamental problems which need to be fully addressed with this regard. These concern: (a) a construction of information granules (both one-dimensional as well as multivariable structures), and (b) exploitation of logic operators and aggregation operators that are carefully adjusted to cope with available experimental data.

In this study, we concentrate on the two design problems identified above and show how they could be efficiently handled by making use of the carefully crafted methodology of fuzzy sets. The design of information granules is discussed in the setting of fuzzy clustering where we envision an incorporation of the machinery of user feedback so that the information granules are formed both on a basis of available experimental evidence (numeric data) whose processing is cast in the framework of a navigation setup formed by the user/designer realized through the formation of the relevance feedback loop. The construction of logic operators aimed at the logic aggregation of information granules builds upon the available data while adhering to the principles of logic computing. Given this character of processing, we will be referring to these constructs as statistically grounded logic aggregators.

## 1 Introductory Remarks

There is a wealth of various formal models of logic connectives in fuzzy sets. There are numerous models which are aimed at revealing an essence of data. These two come hand in hand. By moving from detailed numeric data to information granules, we arrive at a point where data analysis becomes both feasible and user –centric. Fuzzy clustering along with its numerous enhancements helps develop meaningful information granules. Combining information granules calls for efficient logic

E. Rakus-Andersson et al. (Eds.): Recent Advances in Decision Making, SCI 222, pp. 89–101.
springerlink.com 

operators and aggregation mechanisms which are capable of processing a large number of membership degrees. Alluding to the semantics of fundamental logic operators, one can refer to such evident accomplishments in the area as t-norms and t-conorms [6][9], compensative operators [33], aggregative operators [2][14] ordered weighted operators, OWA [15], uninorms [2], and nullnorms. Each of these categories provides additional functionality and in this way offers a highly desirable flexibility to cope with the existing diversity of problems in which fuzzy sets are used. There has been a long way we moved from the introduction of original lattice (min and max) operators on fuzzy sets. In spite of the progress being witnessed in the area, all the pursuits have been predominantly (if not exclusively) motivated by algebraic and logical underpinnings. Surprisingly, not the same amount of attention has been paid to the properties of logic operators and their developments pertinent to handling of numeric experimental data (and membership grades), cf. [33]. These issues are crucial given the need for fostering more advanced and effective techniques of fuzzy modeling. It is needless to say that further advancements in the development of fuzzy systems along with their further applications have posed significant modeling challenges both at the conceptual as well as the optimization end. To address them, there is a definite need for more advanced and computationally plausible logic operators. In particular, parameterized versions of logic operators are of interest as they bring to system modeling the highly desirable flexibility that becomes a genuine necessity when dealing with experimental data. In spite of the number of accomplishments in the realm of the fabric of the logic operators, there are still open questions that deserve careful attention. This concerns issues dealing with a non-pointwise (localized) nature of fuzzy set connectives, cf. [8] and a carefully organized mechanism of incorporation of statistical evidence into logic operators. We envision that such revisited constructs could benefit when being positioned at the junction of logic and the use of the available statistical evidence (results). Our ultimate objective is to consider logic operators whose construction seamlessly embrace the logic fabric and augment it by the existing experimental evidence. Given this, we will be referring to them as statistically grounded OR (SOR) and statistically grounded AND (SAND) logic operators [13].

The objective of this study is to bring the concepts of information granules [4][11][16] and logically and statistically sound aggregation operations to form a unified framework of data analysis and underlying decision-making processes. The general setup we envision involves two main conceptual phases. First, we proceed with a formation of information granules based upon available experimental evidence (numeric data). Such information granules could be formed from a certain fixed perspective of looking at data or several perspectives could be taken into consideration. Second, on a basis of the nature of the information granules obtained during the first phase, the decision-making process is aimed at the discovery of relationships and linkages between information granules. Such dependencies play a pivotal role in the structuralization of knowledge about data. From a more formal perspective, we articulate such findings in terms of overlap between information granules and a level of inclusion between them. Those relationships are expressed by means of possibility and necessity measures and at this point to carry out effective computing of the measures, we engage the use of statistically sound logic operators.

The study is organized in the following manner. We start with a discussion on the design of information granules elaborating on data-driven constricts of fuzzy clustering with a particular focus on the FCM algorithm. In the sequel, in Section 3 we introduce the concept of the statistically grounded logic operators, present the underlying functionality of the constructs and show how SORs and SANDs are constructed as a result of the solutions to a certain optimization problem. The structural data analysis triggers a higher-level data analysis where we inherently exploit some schemes of decision-making processes. The crux of these constructs is discussed in Section 4. Section 5 offers a number of illustrative examples. Throughout the study, we adhere to the commonly utilized notation encountered in fuzzy sets; in particular "t" and "s" will be referred to as t-norms and t-conorms (s-norms), respectively.

## 2 The Design of Information Granules

Fuzzy sets are examples of information granules which are crucial to acquire, organize, and present knowledge about systems under studies. From this perspective, fuzzy clustering offers an interesting and comprehensive insight into the structure of numeric data, cf. [1][3][5][7]13]. Fuzzy clusters form a granular representation of numeric data and therefore constitute their meaningful abstract manifestation, cf.[11]. Let us consider that the results of fuzzy clustering come in the form of "c" clusters built upon a basis of "N" numerical data. Each cluster is described by some fuzzy set $A_i$, i=1, 2, …,c. As a matter of fact, we can envision that $A_i$ forms the i-th row of some partition matrix being the result of the fuzzy clustering. Fuzzy clusters deliver detailed information about the structure in data. The membership grades of individual data to the individual clusters form a useful indicator of their location in the discussed structure. If one of the membership grades is visibly dominant, we may regard the corresponding point to be highly representative for the cluster. On the other hand, if the membership grades of some data are very much equally distributed across all clusters, this sends a strong "flagging" signal as to the borderline character of this point which may eventually trigger further analysis of its properties.

The methods such as Fuzzy C-Means (FCM) give rise to fuzzy sets whose membership functions are determined on a basis of numeric data. They are constructed in a way so that a certain objective function becomes minimized. In the FCM, this objective function is a sum of weighted distances between the data and the prototypes where the weights are the corresponding entries of the partition matrix $U=[u_{ik}]$. More specifically, for the data set $\{\mathbf{x}_1, \mathbf{x}2, \ldots, \mathbf{x}_N\}$, the objective function V reads as follows

$$V=\sum_{i=1}^{c}\sum_{k=1}^{N}u_{ik}^{m}\,||\,\mathbf{x}_k-\mathbf{v}_i\,||^2 \tag{1}$$

where $\mathbf{v}_1, \mathbf{v}_2, \ldots, \mathbf{v}_c$ are the prototypes, m>1 serves as a fuzzification coefficient, and || .|| stands for the distance function between the data and the prototypes. The iterative optimization algorithm is well documented in the literature and the reader may refer to a large number of representative publications.

The FCM algorithm leads to information granules which are exclusively data driven constructs; that is we obtain fuzzy sets on a basis of available numeric data through running some predefined optimization scheme. There are several augmentations of the FCM in which we tend to exploit some knowledge hints provided by humans (users), cf. [11]. The proximity-based fuzzy clustering comes as one of the viable alternatives with this regard. The knowledge hints can come in the form of a collection of so-called “should link” and “should not link” constraints which describe required relationships between some pairs of data. More formally, let us introduce two sets of constraints

$$\Omega = \{\mathbf{x}_k, \mathbf{x}_l \text{ should belong to the same cluster}\} \tag{2}$$

and

$$\Phi = \{\mathbf{x}_k, \mathbf{x}_l \text{ should } \textit{not} \text{ belong to the same cluster}\} \tag{3}$$

The constraints for the pairs of data can be expressed in terms of the proximity matrix which is induced by the partition matrix. More specifically, given the partition matrix (which comes as a result of the fuzzy clustering), the proximity Prox($\mathbf{x}_k$, $\mathbf{x}_l$) is expressed as

$$\text{Prox}(\mathbf{x}_k, \mathbf{x}_l) = \sum_{i=1}^{c} \min(u_{ik}, u_{il}) \tag{4}$$

Given these constraints the objective function becomes augmented and takes the following form

$$V = \sum_{i=1}^{c} \sum_{k=1}^{N} u_{ik}^{m} \| \mathbf{x}_k - \mathbf{v}_i \|^2 + \alpha \sum_{\mathbf{x}_k \mathbf{x}_l \in \Omega} (\text{Prox}(\mathbf{x}_k, \mathbf{x}_l) - 1)^2 \| \mathbf{x}_k - \mathbf{x}_l \|^2 + \beta \sum_{\mathbf{x}_k \mathbf{x}_l \in \Phi} (\text{Prox}(\mathbf{x}_k, \mathbf{x}_l))^2 \| \mathbf{x}_k - \mathbf{x}_l \|^2 \tag{5}$$

The optimization of this objective function is more challenging and calls for the use of advanced optimization mechanisms such as e.g., biologically-inspired optimization (genetic algorithms, particle swarm optimization, etc.)

## 3 Structural Data Analysis

Once the clusters have been constructed through fuzzy clustering, the information granules obtained in this manner can be analyzed in terms of the relationships between them. The two measures which are of interest with this regard are possibility and necessity measure. Let A and B be the two information granules of interest defined over numeric data **X**. In light of our investigations of fuzzy clustering, A and B are treated as some rows of the partition matrix. Let us recall a formal definition of these measures. The two fuzzy sets to be considered are A = [$a_1$ $a_2$ ...$a_N$] and B = [$b_1$ $b_2$ ... $b_N$] defined in **X**. The generalized possibility measure of A and B, Poss(A, B), is defined as follows

$$\text{Poss}(A,B) = \mathop{S}_{i=1}^{N} (a_i t b_i) = \mathop{S}_{i=1}^{N} z_i \tag{6}$$

with $z_i = a_i \, t \, b_i$ where S is a certain t-conorm taken over the successive arguments $z_1$, $z_2$,. $z_N$. The generalization of the possibility measure is sought in terms of the use of some t-conorm in the definition of the measure. In particular, one could consider the maximum operation in which case we end up with a commonly encountered definition of the possibility measure. Given a large number of elements of **X**, we could easily end up with the possibility measure approaching values close to 1.

The generalized necessity measure, Nec(A, B), comes in the form

$$\mathrm{Nec}(A,B) = \mathop{T}_{i=1}^{N}((1-a_i)sb_i) = \mathop{T}_{i=1}^{N} z_i \tag{7}$$

with $z_i = (1-a_i)s\, b_i$ and T being a certain t-norm computed over "N" arguments. In particular, one could envision here the application of the minimum operation returning the "standard" necessity measure. Let us note that the aggregation carried over "N" $x_i$'s very likely leads to the results that converge to zero.

The level of overlap quantified by the possibility measure, Poss (A, B) says how much A and B have in common. The higher the value of the possibility, the more redundancy of A and B is noted: with the increasing possibility measure A is better expressed by B and in this sense become more redundant when dealing with the structural description of data. The necessity measure Nec(A, B) expresses an extent to which A is included in B. In this way it serves as another measure of structural relationship between information granules. This measure is asymmetric so Nec(A, B) ≠ Nec(B, A).

With regard to the analysis of information granules, there are two interesting scenarios whose essence is depicted in Figure 1 in which we emphasize the origin of the information granules.

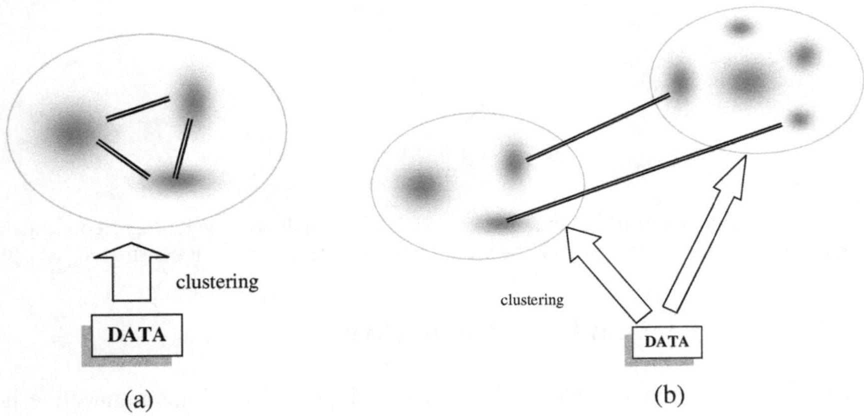


**Fig. 1.** The development of structural relationships between information granules: (a) resulting from the same view at data, and (b) two views at data produced by selected levels of granularity

In the first one, we are interested in expressing relationships between information granules that have been formed from a certain perspective resulting when running the FCM algorithm for a certain predetermined number of clusters. As shown in Figure 1 (b), the relationships are established between information granules which have resulted when considering different levels of granularity through studying different numbers of clusters imposed on the data.

The development of structural relationships could involve more complex constructs. For instance, as visualized in Figure 2, one could have some logic constructs between $A_i$'s for which we are interested to determine relationships with $B_j$'s. In particular one could regard an interesting construct of specialization and abstraction. If the number of clusters card{$A_1, A_2,\ldots, A_c$} is higher than the one encountered in some other view at the data which produces $B_j$'s then the relationships of the form Poss($B_j$, Ai $\bigcup$ $A_k$) could be considered. We say that $B_j$ offers a more general view at the data than the one formed by means of $A_i$s and therefore in expressing $B_j$ in the language of $A_i$'s we utilize a union of the information granules. The dual position is taken when expressing $A_j$'s in terms of $B_k$'s in which an intersection operation is considered, say Poss($A_j$, Bi $\bigcap$ $B_k$). Refer again to Figure 2 for more details. More generally, one could envision the use of some logic expressions in the calculations of the possibility measures, say Poss($\gamma$($A_1$, $A_2$, …, $A_c$), $B_j$) where $\gamma$ stands for this particular logic expression.

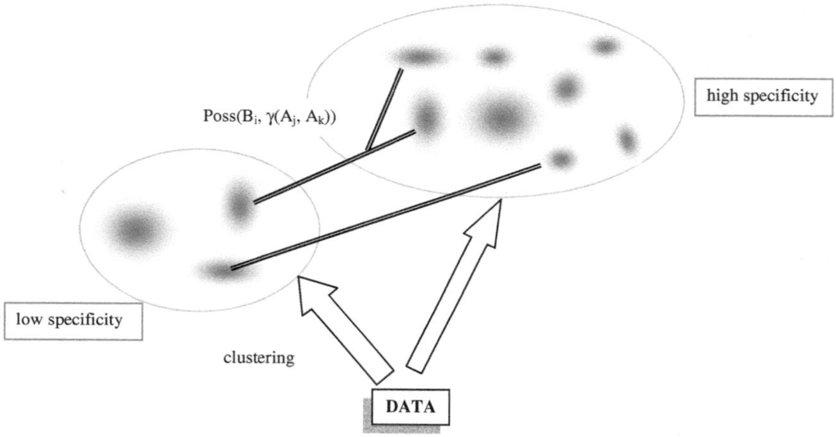


**Fig. 2.** The development of structural relationships in which we invoke logic connectives of intersection and union depending on the granularity level of the granular constructs

## 4 Statistically Sound Logic Connectives

The statistical support incorporated into the structure of the logic connective helps address the issues emerging when dealing with the aggregation schemes articulated by the possibility and necessity measures. We introduce a concept of statistically augmented (directed) logic connectives [12] by constructing a connective that takes into consideration a statistically driven aggregation with some weighting function being reflective of the nature of the underlying logic operation.

### 4.1 SOR Logic Connectives

The (SOR) connective is defined as follows. Denote by w(u) a monotonically non-decreasing weight function from [0,1] to [0,1] with the boundary condition w(1) = 1.

The result of the aggregation of the membership grades $\mathbf{z} = [z_1, z_2, \ldots, z_N]$, denoted by SOR($\mathbf{z}$; w), is obtained as a result of the minimization of the following expression (performance index) Q

$$Q = \sum_{i=1}^{N} w(z_i) \mid z_i - y \mid \qquad \mathrm{Min}_y Q \tag{8}$$

where the value of "y" minimizing the above expression is taken as the result of the operation SOR($\mathbf{z}$, w) = y. Put it differently SOR($\mathbf{z}$, w) = $\arg\min_{y \in [0,1]} \sum_{k=1}^{N} w(z_k) \mid z_k - y \mid$

The weight function "w" is used to model a contribution of different membership grades to the result of the aggregation. Several models of the relationships "w" are of particular interest; all of them are reflective of the *or* type of aggregation

(a) w(z) assumes a form of a certain step function

$$w(z) = \begin{cases} 1 \text{ if } z \geq z_{max} \\ 0, \text{otherwise} \end{cases} \tag{9}$$

where $z_{max}$ is the maximal value reported in $\mathbf{z}$. This weight function effectively eliminates all the membership grades but the largest one. For this form of the weight function, we effectively end up with the maximum operator, SOR($\mathbf{z}$, w) =max ($z_1$, $z_2$, …, $z_N$)

(b) w(z) is equal identically to 1, w(z) =1. It becomes obvious that the result of the minimization of the following expression

$$\sum_{i=1}^{N} \mid z_i - y \mid \tag{10}$$

is a median of $\mathbf{z}$, median($\mathbf{z}$). Subsequently SOR($\mathbf{z}$, w) = median($\mathbf{z}$). Interestingly, the result of the aggregation is a robust statistics of the membership grades involved in this operation.

We can consider different forms of weight functions. In particular, one could think of an identity function w(z) = z. There is an interesting and logically justified alternative which links the weight functions with the logic operator standing behind the logic operations. In essence, the weight function can be induced by various t-conorms (s-norms) by defining w(z) to be in the form w(z) = zsz. In particular, for the maximum operator, we obtain the identity weight function w(z) =max(z, z) = z. For the probabilistic sum, we obtain w(z) = (z+z-z*z) = 2z(1-z). For the Lukasiewicz *or* connective, the weight function comes in the form of some piecewise linear relationship with some saturation region, that is

$$w(z) = \max(1, z+z) = \max(1, 2z).$$

In general, the weight functions (which are monotonically non-decreasing and satisfy the condition w(1)=1) occupy the region of the unit square. For all these weight functions implied by t-conorms, the following inequality holds median(**z**) $\leq$ SOR(**z**, w) $\leq$ max(**z**).

## 4.2 SAND Logic Connectives

The statistically grounded AND (SAND) logic connective is defined in an analogous way as it was proposed in the development of the SOR. Here w(z) denotes a monotonically non-increasing weight function from [0,1] to [0,1] with the boundary condition w(0)=1. The result of the aggregation of $\mathbf{z} = [z_1, z_2, \ldots, z_N]$, denoted by SAND(**z**; w), is obtained from the minimization of the same expression (8) as introduced before. Thus we produce the logic operator SAND(**z**, w) = y with "y" being the solution to the corresponding minimization problem.

As before, we can envision several models of the weight function; all of them are reflective of the *and* type of aggregation

(a) w(z) assumes a form of some step function

$$w(z) = \begin{cases} 1 \text{ if } z \leq z_{min} \\ 0, \text{otherwise} \end{cases} \quad (11)$$

where $z_{min}$ is the minimal value in **z**. This weight function eliminates all the membership grades but the smallest one. For this form of the weight function, we effectively end up with the maximum operator, SAND(**z**, w) =min $(z_1, z_2, \ldots, z_N)$

(b) for w(z) being equal identically to 1, w(z) =1, SAND becomes a median, namely SAND(**z**, w) = med(**z**).

(c) more generally, the weight function is defined on a basis of some t-norm as follows, w(z) =1- ztz. Depending upon the specific t-norm, we arrive at different forms of the mapping. For the minimum operator, w(z) =1- min(z,z) =1-z which is a complement of "z". The use of the product operation leads to the expression w(z) =1-$z^2$. In the case of the Lukasiewicz *and* connective, one has w(z)=1-max(0, z+z-1) =1-max(0, 2z-1).

Investigating the fundamental properties of the logic connectives, we note that the commutativity and monotonicity properties hold. The boundary condition does not hold when being considered with respect to a single membership grade (which is completely understood given the fact that the operation is expressed by taking into consideration a collection of membership grades). Assuming the t-norm and t-conorm driven format of the weight function (where we have w(1) =1 and w(0) =0 for *or* operators and w(0)=1 and w(1)=1 for *and* operators) we have SOR(**1**, w) =0, SAND(**0**, w) = 0. The property of associativity does not hold. This is fully justified given that the proposed operators are inherently associated with the overall processing of all membership grades not just individual membership values.

The possibility and necessity measures determined for the two information granules A and B being articulated in the language of SAR and SAND are expressed in the following manner

$$\text{Poss }(A, B) = \text{SOR}(\mathbf{z}, \mathbf{w}) \; ; \; = a_i t b_i$$
$$\text{Nec }(A, B) = \text{SAND}(\mathbf{z}, \mathbf{w}); \; z_i = a_i s b_i \qquad (12)$$

## 5 Numerical Studies

Several numeric experiments reported in this serve as a illustration of the performance of the method and a way in which the relationships between the information granules are quantified. In all cases, the FCM method uses a "standard" setup: the fuzzification coefficient is equal to 2.0 and the distance is chosen as the Euclidean one (which accounts for substantial differences in the ranges of the individual variables) where the corresponding coordinates were "weighted" by the inverse of their variances. The method was run for 60 iterations. We found that this number was completely sufficient given that no changes in the values of the performance index were reported at this stage. The two data sets used in the experiments available at http://archive.ics.uci.edu/ml/ concern Boston housing and magic data telescope. This second data set has over 19,0020 data points and is one of the largest data set encountered in the Machine Learning repository. When looking at the structure, we reveal and quantify the relationships between information granules by computing the possibility and necessity measures for them. SOR and SAND aggregation operations are realized by making use of the algebraic product (t-norm) and the probabilistic sum (t-conorm). The weight function (w) is implied by the same t-norm and co-norm. Both cases of structural analysis discussed in Section 3, Figure 1 are presented.

Boston housing For the clustering completed for c =3, (with the information granules $A_1$, $A_2$, and $A_3$) the relationships between the information granules are quantified as follows

Possibility measure (consecutive rows and columns correspond to $A_1$, $A_2$, and $A_3$)

$$\begin{bmatrix} & 0.16 & 0.08 \\ 0.16 & & 0.13 \\ 0.08 & 0.13 & \end{bmatrix}$$

Necessity measure

$$\begin{bmatrix} & 0.35 & 0.15 \\ 0.28 & & 0.15 \\ 0.28 & 0.35 & \end{bmatrix}$$

These findings offer an interesting insight into the structural dependencies between the information granules. $A_1$ and $A_2$ come with the highest level of overlap. The lowest overlap occurs for $A_1$ and $A_3$ with the possibility value of 0.08. The necessity values offer another view at the structure by underlying an extent of inclusion occurring between the information granules. Here the highest level is reported for $A_3$ and $A_1$ that is a level of inclusion of $A_3$ in $A_1$. For the calculations of the necessity measure, Figure 3 shows the values of the performance index Q being treated as a function of the optimized result of aggregation; we note that in all cases exhibits a well delineated minimum.

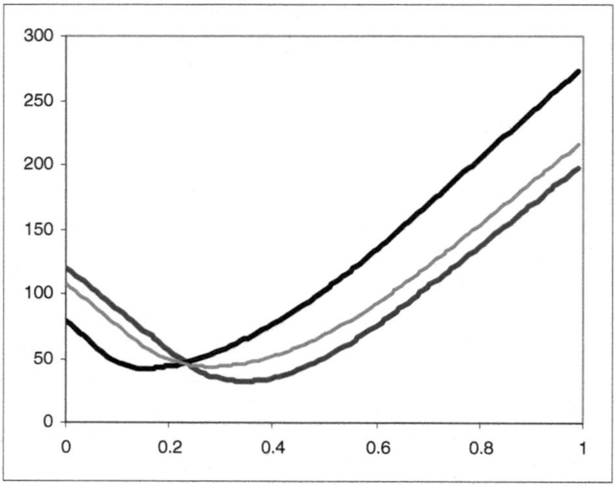


**Fig. 3.** The values of the performance index Q versus the aggregation result

Magic data telescope Here the relationships between the information granules $A_1$, $A_2$, and $A_3$ are quantified in terms of the values of possibility and necessity measures

Possibility

$$\begin{bmatrix} & 0.15 & 0.14 \\ 0.15 & & 0.07 \\ 0.14 & 0.07 & \end{bmatrix}$$

Necessity

$$\begin{bmatrix} & 0.17 & 0.24 \\ 0.33 & & 0.24 \\ 0.33 & 0.17 & \end{bmatrix}$$

Considering that we have developed a family of information granules for different values of c, the relationships (possibility values) between them are quantified in the following matrices containing the corresponding possibility values,

(2, 5)

$$\begin{bmatrix} 0.20 & 0.10 \\ 0.07 & 0.18 \\ 0.32 & 0.07 \\ 0.07 & 0.18 \\ 0.14 & 0.14 \end{bmatrix}$$

(3, 6)

$$\begin{bmatrix} 0.09 & 0.04 & 0.15 \\ 0.11 & 0.06 & 0.09 \\ 0.06 & 0.04 & 0.26 \\ 0.07 & 0.11 & 0.04 \\ 0.09 & 0.09 & 0.05 \\ 0.07 & 0.11 & 0.04 \end{bmatrix}$$

These relationships offer and interesting view at the information granules. For the pair of (2, 5) information granules, we learn that the highest overlap occurs between $A_3$ and $B_1$ (which stipulates that there is the highest level of conceptual redundancy between these two information granules). On the other hand, the linkages between $A_4$ and $B_1$ as well as $A_3$ and $B_2$ are very weak and this points at the lower level of redundancy concerning these particular granules. The low values of the possibility measure for the information granules mean that they represent quite disjoint regions of data as far as the underlying structure is concerned. In other words, more specific (detailed) information granules bring more structural insights that have not been already captured when working with information granules of lower specificity (viz. higher abstraction). For the second pair of information granules (3, 6) we have the highest overlap (viz. redundancy) for $A_1$ and $B_3$ and the lowest one equal to 0.04 occurs for $A_4$ and $B_3$.

For the second simulation scenario with the pair of 6 and 3 clusters, the most significant overlap (redundancy) between information granules occurs in case of $A_3$ and $B_3$ (where the possibility value is 0.26) and $A_1$ and $B_3$ (with the possibility value equal to 0.15). One can take another general look at the relationship between Ai and Bj: as the granularity of $B_j$'s is lower than those $A_i$'s, it is likely that some $A_i$'s are relatively well represented by $B_j$'s. The results shown in the matrix of the possibility values capture this effect. As noted earlier, $B_3$ is representative of $A_1$ and $A_3$. For $B_1$, it represents $A_2$ (with the possibility value equal to 0.11). The representation of $B_2$ is mainly associated with $A_4$ and $A_6$ (with the possibility level of 0.11).

## 6 Conclusions

Being fully cognizant of the challenges of fuzzy data analysis carried out at the level of information granules, by proposing statistically grounded logic operators, we have emphasized the need for more data driven – constructs that dwell on available experimental evidence. We have developed logic operators that take into consideration collections of numeric membership grades and exploit their statistical characteristics through the use of the weight function. Interestingly, the weight function underlines the logic nature of the operator. The OR class of logic operators, named here SOR, is generated by the weight functions that are monotonically nondecreasing functions and constructed by involving some t-norm, w(u) =utu or more generally w(u) =g(usu) with "g" being a certain monotonically nondecreasing mapping. The category of statistically grounded AND operators, SAND, is generated by the weight functions that are monotonically nonincreasing over the unit interval. We discussed the weight functions of the form w(u) =1- (utu); the general form of the relationship could be sought as w(u) = h(1-(utu)) with "h" being a monotonically nondecreasing mapping on the unit interval. The choice of the t-norm or t-conorm used in the SOR or SAND could be treated as a part of the design process: given some data that are to be approximated by the logic operator, we can choose a suitable triangular norm (conorm) in the weight function so that the best approximation (viz. with the lowest approximation error) is achieved.

The statistically sound logic operators constitute a cornerstone of the possibility and necessity measures using which we reveal and quantify the relationships between information granules. We showed that when dealing with a large number of arguments (membership grades), the statistical nature of the collections of membership grades becomes critical and need to be incorporated as a part of the underlying construct.

The study can be sought as a first attempt to bring statistically sound operators into the realm of data analysis. Interestingly, we could envision a number of further directions worth exploring. In particular, one can investigate ways of reconciliation/aggregation of granular findings supported by individual sources of numeric data.

## Acknowledgments

Support from the Natural Sciences and Engineering Research Council of Canada (NSERC) and Canada Research Chair (CRC) is gratefully acknowledged.

## References

1. Bezdek, J.C.: Pattern Recognition with Fuzzy Objective Function Algorithms. Plenum Press, New York (1981)
2. Calvo, T., Kolesárová, A., Komorníková, M., Mesiar, R.: Aggregation operators: properties, classes and construction methods. In: Aggregation Operators: New Trends and Applications, pp. 1–104. Physica-Verlag, Heidelberg (2002)
3. Cardoso, G., Gomide, F.: Newspaper demand prediction and replacement model based on fuzzy clustering and rules. Information Sciences 177(21), 4799–4809 (2007)
4. Kacprzyk, J., Zadrozny, S.: Linguistic database summaries and their protoforms: towards natural language based knowledge discovery tools. Information Sciences 173(4), 281–304 (2005)
5. Kiliç, K., Uncu, O., Türksen, I.B.: Comparison of different strategies of utilizing fuzzy clustering in structure identification. Information Sciences 177(23), 5153–5162 (2007)
6. Klement, E., Mesiar, R., Pap, E.: Triangular Norms. Kluwer Academic Publishers, Dordrecht (2000)
7. Ozkan, I., Türkşen, I.B., Canpolat, N.: A currency crisis and its perception with fuzzy C-means. Information Sciences 178(8), 1923–1934 (2008)
8. Pedrycz, W.: Associations of fuzzy sets. IEEE Trans. on Systems, Man and Cybernetics 6, 1483–1488 (1992)

9. Pedrycz, W., Gomide, F.: An Introduction to Fuzzy Sets: Analysis and Design. MIT Press, Cambridge (1998)
10. Pedrycz, W.: Fuzzy clustering with a knowledge-based guidance. Pattern Recognition Letters 25(4), 469–480 (2004)
11. Pedrycz, W.: Knowledge-Based Clustering. J. Wiley, Hoboken (2005)
12. Pedrycz, W.: Statistically grounded logic operators in fuzzy sets. European Journal of Operational Research 193(2), 520–529 (2009)
13. Tari, L., Baral, C., Kim, S.: Fuzzy c-means clustering with prior biological knowledge. Journal of Biomedical Informatics (online May 24, 2008) (in press) (corrected proof)
14. Torra, V.: Aggregation operators and models. Fuzzy Sets and Systems 156, 407–410 (2005)
15. Yager, R.: On ordered weighted averaging aggregation operations in multicriteria decision making. IEEE Trans. on Systems, Man, and Cybernetics 18, 183–190 (1988)
16. Zadeh, L.A.: Toward a generalized theory of uncertainty (GTU) -— an outline. Information Sciences 172, 1–40 (2005)

# 7

# A General Methodology for Managerial Decision Making Using Intelligent Techniques*

Vilém Novák[1], Irina Perfilieva[1], and Nadezhda G. Jarushkina[2]

[1] University of Ostrava
Institute for Research and Applications of Fuzzy Modeling
30. dubna 22, 701 03 Ostrava 1, Czech Republic
{Vilem.Novak,Irina.Perfilieva}@osu.cz
[2] Ulyanovsk State Technical University
Severny venec, 32
432027 Ulyanovsk, Russia
jng@ultsu.ru

**Abstract.** Managerial decision-making is a complex procedure which combines information both in numerical as well as in linguistic form. In this paper, we present a general decision-making methodology which utilizes several intelligent soft-computing techniques, namely the theory of evaluative linguistic expressions, perception-based logical deduction and fuzzy transform. These techniques fulfil the above requirement and so, we are convinced that they can effectively fulfil the needs of managers and provide them with a tool that can help them to obtain a relevant decision. The methodology is demonstrated on an example.

**Keywords:** Soft computing, Fuzzy logic in broader sense, Fuzzy decision-making, Evaluative linguistic expressions, Fuzzy transform.

## 1 Introduction

Managerial decision-making is a complex procedure which combines information both in numerical form as well as in linguistic form. Natural language, however, is subjected to imprecision and vagueness. Therefore, managers are somewhat reluctant to using mathematical methods for decision support due to their limited applicability in practice. In this paper, we present a general methodology which combines several intelligent techniques capable at elaboration of both kinds of information so that a relevant decision can be reached. The techniques are well suitable for managerial decision-making because they can provide optimal decisions close to decisions made by people.

Let us remark that there are already thousands of publications on decision making, and also of those devoted to the fuzzy multicriteria one. These works have been initiated by the paper [1]. Among many books on this topic, let us mention, for example [5, 18], or recently [7].

* The paper has been supported by the project 1M0572 of the MŠMT ČR.

E. Rakus-Andersson et al. (Eds.): Recent Advances in Decision Making, SCI 222, pp. 103–120.
springerlink.com 

One of typical problems raising in decision making stems from the fact that the considered criteria are usually not equally important. Their relative importance is expressed using weights. A popular method for weights assignment is *Analytical Hierarchy Process* (AHP) that itself can be used as a specific fuzzy decision method [16]. It involves structuring multiple choice criteria into a hierarchy, assessing the relative importance of these criteria, comparing alternatives for each criterion, and determining an overall ranking of the alternatives. Other important problem is aggregation of criteria to obtain final decision. In fuzzy approach, this is achieved using aggregation operators (cf. [3]).

This paper differs from the above mentioned works in some respects, because it combines special soft-computing techniques to obtain a realistic decision:

- Techniques developed in the frame of fuzzy logic in broader sense. These are based on a mathematical model of the semantics of a part of natural of natural language, which covers evaluative linguistic expressions and intermediate quantifiers.
- A special inference method called *perception-based logical deduction* (PbLD) using which a conclusion on the basis of linguistic description of the given decision situation can be obtained. The linguistic description consists, in general, of rules being conditional clauses of natural language.
- A special soft computing technique called *fuzzy transform* (F-transform), which is a general mathematical technique for robust approximation of functions. One of many applications of it is analysis and prediction of time series. For managers it is most important to estimate the principal course and tendency of time series to be able to find decisions having strategic character. The F-transform in combination with PbLD is very powerful for this purpose.

The outcome of these methods is twofold: first, imprecise information contained in natural language is effectively utilized. Second, some problems otherwise necessary to be solved, such as assignment of weights and aggregation of criteria, are replaced, in our opinion, by more natural techniques.

Our decision-making methodology leads to classical preference relation, i.e. the alternatives are linearly ordered on the basis of evaluation that behaves as special utility function.

In this paper, we will briefly describe the mentioned techniques from the point of view of decision making. Furthermore, we develop a general decision-making methodology using them and demonstrate the methodology on a sophisticated example. At the end, we mention elaboration of time series and demonstrate what kind of information can be obtained and how it can be used in managerial decision-making.

## 2 Basic Soft Computing Techniques

We will identify fuzzy sets on a universe $U$ with their membership functions, i.e., a fuzzy set is a mapping from $U$ into $[0,1]$. If $A$ is a fuzzy set in the universe $U$

the we often write $A \underset{\sim}{\subseteq} U$. The set of all fuzzy sets over a universe $U$ is denoted by $\mathscr{F}(U)$. By $\mathbb{R}$ we denote a set of real numbers.

### 2.1 Evaluative Linguistic Expressions and Linguistic Description

Evaluative linguistic expressions are special expressions of natural language that are used whenever it is important to evaluate a decision situation, a course of development of some process, characterize manifestation of some property, and in many other specific situations. Typical examples of evaluative (linguistic) expressions are "very large, extremely expensive, roughly one thousand, more or less hot", etc. Note that their importance and the proposal to model mathematically their meaning has been pointed out by L. A. Zadeh (e.g., [19–21] and elsewhere). A formal theory of them is elaborated in detail in [9]. It includes a mathematical model of their semantics, which is considered also in this paper.

We will deal with simple form of evaluative expressions having the following syntactic structure:

$$\langle\text{linguistic hedge}\rangle\langle\text{atomic evaluative expression}\rangle \tag{1}$$

or

$$\langle\text{linguistic hedge}\rangle\langle\text{numeral}\rangle \tag{2}$$

where "numeral" is a name of some number $x_0 \in \mathbb{R}$. *Atomic evaluative expressions* and *numerals* form the basic component of all kinds of evaluative expressions. Atomic expressions comprise any of the *canonical* adjectives *small, medium, big*. It is important to stress that these words are in practice often replaced by other kinds of evaluative words such as "thin", "thick", "old", "new", etc., depending on the context of speech.

Linguistic hedges are specific adverbs which make the meaning of the atomic expression more or less precise. We may classify hedges to those with *narrowing effect*, for example *very, significantly, extremely*, etc., those with *widening effect*, for example *roughly, more or less, quite roughly*, etc. and *mixed* ones, for example *rather, approximately*, etc. The evaluative expressions of the form (1) will generally be denoted by $\mathrm{Ev}_\nu$ where $\nu$ is the linguistic hedge. Note that as a special case, the ⟨linguistic hedge⟩ can be empty. This enables us to identify atomic evaluative expressions with simple ones and develop a unified theory of their meaning.

Evaluative expressions are used for evaluation of values of some variable $X$. The resulting expressions are called *evaluative (linguistic) predications* and have the form

$$X \text{ is } \mathrm{Ev}_\nu\,. \tag{3}$$

Examples of evaluative predications are "temperature is very high", "price is low", "pressure is rather strong", etc.).

Our model makes distinction between *intension* of an evaluative predication and its *extensions* in various *contexts*. The context characterizes a range of possible values and is determined by a triple $\langle v_L, v_M, v_R\rangle$, where $v_L, v_M, v_R \in \mathbb{R}$

and $v_L < v_M < v_R$ characterize minimal, middle and maximal value of the given context, respectively. By $u \in w$ we mean $u \in [v_L, v_R]$. We suppose to be given a set of contexts $W$,

$$W = \{\langle v_L, v_M, v_R\rangle \mid v_L, v_M, v_R \in \mathbb{R}, v_L < v_M < v_R\}.$$

*Intension* of an evaluative predication "$X$ is $\mathrm{Ev}_\nu$" is a certain formula (cf. [9]) whose interpretation is a function

$$\mathrm{Int}(X \text{ is } \mathrm{Ev}_\nu) : W \longrightarrow \mathscr{F}(\mathbb{R}). \tag{4}$$

Given an intension of an evaluative predication and a context $w \in W$, we can define its *extension* as a fuzzy set

$$\mathrm{Int}(X \text{ is } \mathrm{Ev}_\nu)(w) \underset{\sim}{\subset} [v_L, v_R]$$

where $v_L, v_R$ are left and right bound of the given context $w = \langle v_L, v_M, v_R\rangle$.

Evaluative predications occur in linguistic conditional rules

$$\mathscr{R} := \mathsf{IF}\ X \text{ is } \mathrm{Ev}_\nu^X\ \ \mathsf{THEN}\ Y \text{ is } \mathrm{Ev}_\nu^Y\,. \tag{5}$$

where $\mathrm{Ev}_\nu^X, \mathrm{Ev}_\nu^Y$, $j = 1, \ldots, m$ are evaluative expressions. The linguistic predication "$X$ is $\mathrm{Ev}_\nu^X$" is called *antecedent* and "$Y$ is $\mathrm{Ev}_\nu^Y$" is called *consequent* of the rule (5). Of course, the antecedent may consist of more evaluative predications which, when present, are joined by the connective "AND". We will call (5) a fuzzy IF-THEN rule*) in the sequel.

*Intension of a fuzzy IF-THEN rule* $\mathscr{R}$ in (5) is a function

$$\mathrm{Int}(\mathscr{R}) : W \times W \longrightarrow \mathscr{F}(\mathbb{R} \times \mathbb{R}). \tag{6}$$

This function assigns to each context $w \in W$ and each context $w' \in W$ a *fuzzy relation* in $w \times w'$. The latter is a *extension of the rule* (6).

Fuzzy IF-THEN rules are gathered in a *linguistic description* which is a set $LD = \{\mathscr{R}_1, \ldots, \mathscr{R}_m\}$ of fuzzy IF-THEN rules:

$$\begin{aligned} \mathscr{R}_1 &= \mathsf{IF}\ X \text{ is } \mathrm{Ev}_{\nu,1}^X\ \ \mathsf{THEN}\ Y \text{ is } \mathrm{Ev}_{\nu,1}^Y, \\ &\cdots\cdots\cdots\cdots\cdots\cdots\cdots\cdots \\ \mathscr{R}_m &= \mathsf{IF}\ X \text{ is } \mathrm{Ev}_{\nu,m}^X\ \ \mathsf{THEN}\ Y \text{ is } \mathrm{Ev}_{\nu,m}^Y\,. \end{aligned} \tag{7}$$

Because each rule in (7) is taken as a specific *conditional sentence of natural language*, a linguistic description can be understood as a *specific kind of a (structured) text.*

We will also need to consider a linguistic phenomenon of topic-focus articulation (cf. [6, 17]), which in case of fuzzy IF-THEN rules requires to distinguish the following two sets:

$$\begin{aligned} \mathit{Topic}_{LD} &= \{\mathrm{Int}(X \text{ is } \mathrm{Ev}_{\nu,j}^X) \mid j = 1, \ldots, m\}, \\ \mathit{Focus}_{LD} &= \{\mathrm{Int}(Y \text{ is } \mathrm{Ev}_{\nu,j}^Y) \mid j = 1, \ldots, m\}. \end{aligned}$$

The phenomenon of topic-focus articulation plays an important role in the inference method called perception-based logical deduction described below.

*) More precisely, we should call it fuzzy/linguistic IF-THEN rule.

### 2.2 Perception, Evaluation and Learning

By perception we will understand an evaluative expression assigned to the given value in the given context. The choice of perception is not arbitrary and it also depends on the topic of the specified linguistic expression.

We define a special function of *local perception*

$$LPerc^{LD} : w \times W \longrightarrow Topic_{LD} \tag{8}$$

assigning to each value $u \in w$ for $w \in W$ an intension

$$LPerc^{LD}(u, w) = \mathrm{Int}(X \text{ is } \mathrm{Ev}^X_{\nu,j}) \tag{9}$$

of the *sharpest* evaluative predication (w.r.t. the specific ordering) so that $u \in w$ is the *most typical* element for the extension $\mathrm{Int}(X \text{ is } \mathrm{Ev}^X_{\nu,j})(w)$. If there is no evaluative expression being most specific and typical then (9) is undefined.

The concept of perception is connected with the concept of evaluation. We say that *an element $u \in w$ is evaluated by an evaluative expression* $\mathrm{Ev}_\nu$, if there is a truth value $a \neq 0$ and $u \in w$ such that $a \rightarrow (\mathrm{Int}(X \text{ is } \mathrm{Ev}_\nu)(w))(u) = 1$ where $\rightarrow$ is a fuzzy implication function (this is usually the Łukasiewicz implication*)). Then we write formally

$$Eval(u, w, \mathrm{Ev}_\nu). \tag{10}$$

The idea of assigning local perception (9) needs not be restricted only to the topic. If we slightly generalize it, we can learn the linguistic description on the basis of the given data. More details about the learning method can be found in [2]. Let us remark that we have successfully implemented this method in the software system LFLC2000 and applied it to the forecasting of time series.

### 2.3 Perception-Based Logical Deduction

Let us be given a linguistic description $LD$ in (7), a context $w \in W$ for the variable $X$ and a context $w' \in W$ for $Y$. Furthermore, let an observation $X = u_0$ in the context $w$ be given where $u_0 \in w$. Using (9), we assign $u_0$ a perception $\mathrm{Int}(X \text{ is } \mathrm{Ev}^X_{\nu,j_0})$.

On the basis of that the following *rule of perception-based deduction* is valid:

$$r_{PbLD} : \frac{LPerc^{LD}(u_0, w) \equiv \mathrm{Int}(X \text{ is } \mathrm{Ev}^X_{\nu,j_0}), \quad LD}{Eval(\hat{v}, w', \mathscr{B}_i)} \tag{11}$$

where $\mathrm{Int}(X \text{ is } \mathrm{Ev}^X_{\nu,j_0}) \in Topic^{LD}$, $\mathrm{Int}(X \text{ is } \mathrm{Ev}^Y_{\nu,j_0}) \in Focus^{LD}$. Practically useful result is obtained after interpretation of (11). Therefore, we will denote a truth value

$$C(v) = (\mathrm{Int}(X \text{ is } \mathrm{Ev}^X_{\nu,j_0})(w))(u_0) \rightarrow (\mathrm{Int}(Y \text{ is } \mathrm{Ev}^Y_{\nu,j_0})(w'))(v).$$

*) Łukasiewicz implication is in $[0, 1]$ defined by the formula $a \rightarrow b = \min\{1, 1 - a + b\}$.

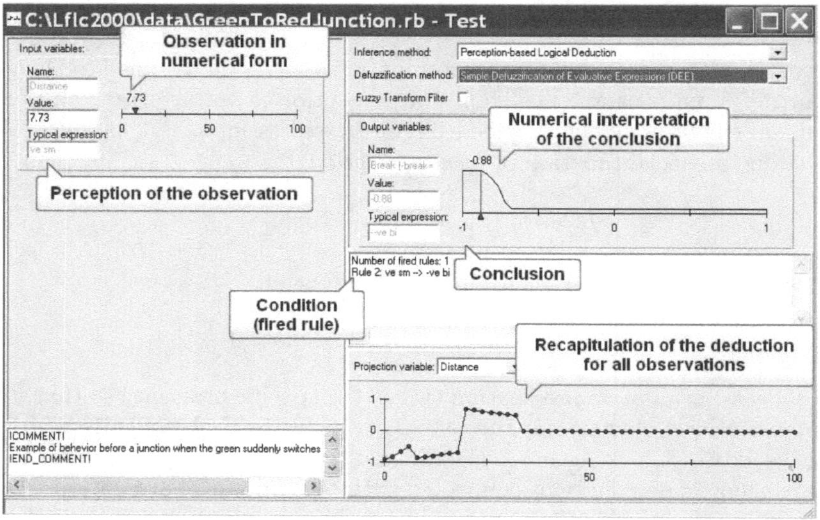


**Fig. 1.** The work of perception-based logical deduction

(recall that the antecedent $(\mathrm{Int}(X \text{ is } \mathrm{Ev}^X_{\nu,j_0})(w))(u_0)$ is a membership degree of the observation $u_0$ in the extension being a fuzzy set $\mathrm{Int}(X \text{ is } \mathrm{Ev}^X_{\nu,j_0})(w)$ and similarly also the consequent).

Then the result of PbLD is an element $\hat{v} \in w'$ given by

$$\hat{v} = \mathrm{DEE}\left(\left\{C(v)/v \mid v \in w'\right\}\right) \tag{12}$$

where DEE is a special defuzzification method (Defuzzification of Evaluative Expressions).

The PbLD method can be in free words described as follows: if a linguistic description (7) consisting of fuzzy/linguistic IF-THEN rules together with an observation of some value of the variable $X$ are given then the PbLD method chooses the most proper rule with respect to the topic of the linguistic description and derives a typical element (12) from its conclusion. The act of the procedure together with explanation is demonstrated in Figure 1. Detailed formal analysis of PbLD as well its justification can be found in [8, 10].

### 2.4 Fuzzy Transform

The fuzzy transform (F-transform) is a technique developed by I. Perfilieva [14] which can be ranked among *fuzzy approximation techniques*. Its basic idea is to transform a continuous function $f$ defined on an interval of real numbers $w = [v_L, v_R] \subset \mathbb{R}$ into a vector of numbers $[F_1, \ldots, F_n]$ (we speak about direct F-transform). After realizing operations with this vector, the result can be transformed back to the original space (we speak about inverse F-transform). For practical applications, it is useful to suppose that the function $f$ is determined in finite number of points. In this case, we speak about *discrete* F-transform.

The F-transform is a general technique having many applications. Besides approximation of $f$ with arbitrary precision, F-transform can be applied for filtering of noise, analysis of time series, image processing, and in many other applications.

**Fuzzy Partition**

The principal concept in the theory of F-transform is that of a fuzzy partition. Let us suppose that the function $f$ is known at *points* $p_1, \ldots, p_N \in w$. We divide the interval $w = [v_L, v_R]$ into a set of equidistant *nodes* $x_k = v_L + h(k-1)$, $k = 1, \ldots, n$ where $N > n$, $h = \frac{v_R - v_L}{n-1}$ is a fixed length. Obviously, $x_1 = v_L$ and $x_n = v_R$.

We say that fuzzy sets $A_1, \ldots, A_n \underset{\sim}{\subset} [v_L, v_R]$ constitute a *fuzzy partition* of $[v_L, v_R]$ if they fulfill the following conditions for $k = 1, \ldots, n$:

(i) $A_k : [v_L, v_R] \longrightarrow [0, 1]$, $A_k(x_k) = 1$;
(ii) $A_k(x) = 0$ if $x \notin (x_{k-1}, x_{k+1})$ where for the uniformity of denotation, we put $x_0 = v_L$ and $x_{n+1} = v_R$;
(iii) $A_k(x)$ is continuous;
(iv) $A_k(x)$, $k=2, \ldots, n$, strictly increases on $[x_{k-1}, x_k]$ and $A_k(x)$, $k=1, \ldots, n-1$, strictly decreases on $[x_k, x_{k+1}]$;
(v) $\sum_{k=1}^{n} A_k(x) = 1, \qquad x \in [a, b]$.

The membership functions $A_1(x), \ldots, A_n(x)$ are called *basic functions.*

Let us remark that basic functions are specified by a set of nodes $x_1 < \ldots < x_n$ and the properties (i)–(v). The shape of basic functions is not predetermined and therefore, it can be chosen additionally according to further requirements (e.g., smoothness).

We say that the fuzzy partition $A_1, \ldots, A_n$, $n \geq 3$, is *uniform* if the nodes $x_1, \ldots, x_n$ are equidistant, i.e. $x_k = v_L + h(k-1)$ where $h = (v_R - v_L)/(n-1)$, and two additional properties are fulfilled:

(vi) $A_k(x_k - x) = A_k(x_k + x)$, for all $x \in [0, h]$, $k = 2, \ldots, n-1$,
(vii) $A_k(x) = A_{k-1}(x - h)$, for all $k = 2, \ldots, n-1$ and $x \in [x_k, x_{k+1}]$, and $A_{k+1}(x) = A_k(x - h)$, for all $k = 2, \ldots, n-1$ and $x \in [x_k, x_{k+1}]$.

**The Fuzzy Transform Technique**

We will assume that $v_L = x_1 < \ldots < x_n = v_R$ are fixed nodes and $p_1, \ldots, p_l \in [v_L, v_R]$ are fixed points such that $n \geq 2$, $l > n$. Let $A_1, \ldots, A_n$ be a fuzzy partition of $[v_L, v_R]$. We must also assume that the set $P_l = \{p_1, \ldots, p_l\}$ is *sufficiently dense with respect to the partition,* i.e. that

$$(\forall k)(\exists j) \quad A_k(p_j) > 0 \tag{13}$$

holds true. Then the F-transform has two phases.

*Direct F-transform*

The function $f$ is transformed into $n$-tuple of real numbers $[F_1, \ldots, F_n]$ defined by

$$F_k = \frac{\sum_{j=1}^{N} f(p_j)A_k(p_j)}{\sum_{j=1}^{N} A_k(p_j)}, \qquad k = 1, \ldots, n. \tag{14}$$

Each number $F_k$, $k = 1, \ldots, n$ is called a *component* of fuzzy transform.

*Inverse F-transform.* The vector of numbers $[F_1, \ldots, F_n]$ contains information about the original function $f$. Therefore, it can be used to obtain a function

$$f_{F,n}(x) = \sum_{k=1}^{n} F_k \cdot A_k(x). \tag{15}$$

The function $f_{F,n}(x)$ is called the *inverse F-transform* of $f$. It can be proved that if $n$ increases then $f_{F,n}(p_j)$ converges to $f(p_j)$, $j = 1, \ldots, N$.

The F-transform technique is schematically depicted in Figure 2. It has (besides others) the following properties:

(a) It has nice filtering properties.
(b) It is easy to compute.
(c) The F-transform is stable with respect to the choice of the points $p_1, \ldots, p_N$. This means that when choosing other points $p_k$ (and possibly changing their number $N$), the resulting function $f_{F,n}$ does not significantly change. Note that this is not true for many classical numerical methods.

A detailed formal description of the F-transform including many theorems can be found in[13] (see also [12]).

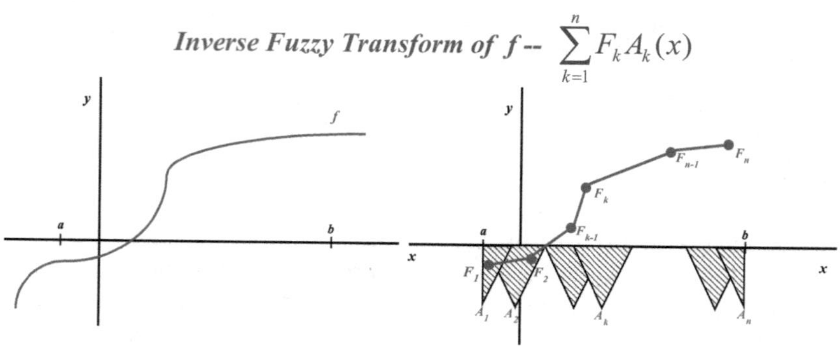


**Fig. 2.** Scheme of F-transform. The left-hand side shows the original function $f$. The right-hand depicts the fuzzy partition, the corresponding F-transform components and the approximation of $f$ being a result inverse F-transform.

# 3 Soft-Computing Techniques in Decision Making

## 3.1 Decision Making Using Perception-Based Logical Deduction

The assumption that fuzzy IF-THEN rules are linguistically described logical implications characterizing relation among vaguely characterized phenomena makes it possible to distinguish sufficiently subtly and, at the same time, aptly, various degrees of fulfilment of the respective criteria, distinguish their importance and, moreover, overcome possible discrepancies. If the decision situation is sufficiently well characterized by linguistic descriptions then the proper decision can be done using PbLD.

It is important to note that the problem of assignment of weights to the criteria disappears. This is one of the great advantages of our approach because methods for weights assignment, though sophisticated, are rather intricate and still much subjective (though the weights are seemingly objective). Of course, formulation of the problem using fuzzy IF-THEN rules is also subjective. However, this subjectivity is here under control because it is formulated using natural language understandable to everybody. Therefore, people (experts) may agree on the optimal formulation thus reflecting the general knowledge.

The multicriteria decision-making problem can be described as follows. Let $n$ criteria $C_1, \ldots, C_n$ be given. On the basis of them, we should decide about alternatives $v_1, \ldots, v_m$. Fulfilling of each criterion is measured by values within a specific scale. This scale determines the above considered linguistic context. There are two possibilities:

(i) The given criterion is measurable by real measuring units. This is the case where a criterion has an objective character such as price, geometrical or physical characteristics, etc. The scale (linguistic context) is then determined by specific application (for example, economical power of the given company, concrete object, etc.)
(ii) The criterion is subjective or abstract but people can estimate how well it is fulfilled by a given alternative. The scale is then usually set as the abstract interval $[0, 1]$ of spaceless units. The corresponding linguistic context is set to $w = \langle 0, 0.4, 1 \rangle$.

Using fuzzy IF-THEN rules, it is possible to characterize clearly how the criteria should be fulfilled so that the global characterization of the given alternative were as best as possible. This practically means that the decision-making situation is described using a linguistic description consisting of rules of the form

$$\mathscr{R}_1 := \text{IF } C_1 \text{ is } \mathscr{A}_{11} \text{ AND } \ldots \text{ AND } C_n \text{ is } \mathscr{A}_{1n} \text{ THEN } H \text{ is } \mathscr{B}_1$$

$$\ldots\ldots\ldots\ldots\ldots\ldots\ldots\ldots\ldots\ldots\ldots\ldots\ldots\ldots\ldots$$

$$\mathscr{R}_p := \text{IF } C_1 \text{ is } \mathscr{A}_{p1} \text{ AND } \ldots \text{ AND } C_n \text{ is } \mathscr{A}_{pn} \text{ THEN } H \text{ is } \mathscr{B}_p$$

where $C_1, \ldots, C_n$ are the criteria, $H$ is global evaluation and $\mathscr{A}_{ji}, \mathscr{B}_{ji}$, $j = 1, \ldots, m$, $i = 1, \ldots, p$ are the above discussed evaluative linguistic expressions.

However, the number $n$ can be large and so, it may not be possible to form such a linguistic description (it could even be hardly understandable). Therefore, we will formally divide the criteria $C_1, \ldots, C_n$ into $r$ groups $H_1, \ldots, H_r$. Let us denote

$$\mathscr{H}_k = \{C_{k1}, \ldots, C_{kn(k)}\}, \qquad k = 1, \ldots, r.$$

Then the above introduced linguistic description can be transformed into a hierarchical system of linguistic descriptions

$$\mathscr{R}_{1k} := \text{IF } C_{k1} \text{ is } \mathscr{A}_{11} \text{ AND } \ldots \text{ AND } C_{kn(k)} \text{ is } \mathscr{A}_{1n(k)} \text{ THEN } H_k \text{ is } \mathscr{B}_1$$

$$\ldots\ldots\ldots\ldots\ldots\ldots\ldots\ldots\ldots\ldots\ldots\ldots\ldots\ldots\ldots$$

$$\mathscr{R}_{1p(k)} := \text{IF } C_{k1} \text{ is } \mathscr{A}_{p(k)1} \text{ AND } \ldots \text{ AND } C_{kn(k)} \text{ is } \mathscr{A}_{p(k)n(k)} \text{ THEN } H_k \text{ is } \mathscr{B}_{p(k)}$$

$k = 1, \ldots, r$ and

$$\mathscr{R}_1 := \text{IF } H_1 \text{ is } \mathscr{A}_{11} \text{ AND } \ldots \text{ AND } H_r \text{ is } \mathscr{A}_{1r} \text{ THEN } H \text{ is } \mathscr{B}_1$$

$$\ldots\ldots\ldots\ldots\ldots\ldots\ldots\ldots\ldots\ldots\ldots\ldots\ldots\ldots\ldots$$

$$\mathscr{R}_s := \text{IF } H_1 \text{ is } \mathscr{A}_{s1} \text{ AND } \ldots \text{ AND } H_r \text{ is } \mathscr{A}_{sr} \text{ THEN } H \text{ is } \mathscr{B}_s.$$

The last linguistic description provides a final evaluation using which the proper decision is made.

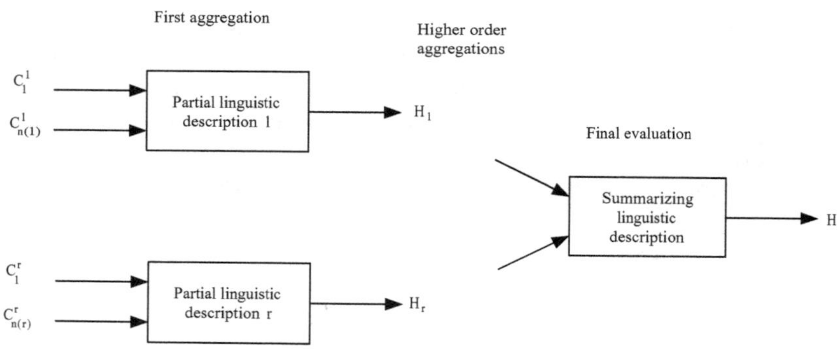


**Fig. 3.** General scheme of the hiearchy of decision-making using linguistic descriptions

### 3.2 Demonstration of the Decision-Making Methodology

Let us demonstrate the use of soft computing techniques on a model of a complex decision-making problem similar to problems raising in the reality. The basic tool is the software system LFLC 2000 developed at the University of Ostrava (cf. [4]).

Our decision problem consists in choosing one of 20 houses offered by a one real estate agency of the Czech Republic on internet. For each house, various kinds of information are available (including its photo).

**Table 1.** Input data of four selected houses. The linguistic expressions are coded using shorts with the following meaning: *sm*–small, *me*–medium, *bi*–big, *ex*-extremely, *si*–significantly, *ve*–very, *ml*–more or less, *qr*–quite roughly, *ro*–roughly, *ra*–rather, *no*–not.

| | Econ. Char | | Tech. Char | | | | Sizes | |
|---|---|---|---|---|---|---|---|---|
| House | Price 1000CzK | Recnstr. cost | No. rooms | No. bath | No. WC | Heating | House size | Land size |
| 1. | 2500 | bi | 4 | 1 | 1 | ra bi | 7200 | 0 |
| 6. | 2690 | ve sm | 5 | 1 | 2 | me | 1141 | 961 |
| 12. | 3300 | ml bi | 7 | 3 | 3 | bi | 1800 | 1600 |
| 19. | 3999 | ex sm | 8 | 3 | 2 | ra bi | 855 | 505 |

| Other Char | | Infrastructure | | | Aest. char. | | | |
|---|---|---|---|---|---|---|---|---|
| Garage | Cellar | Dist center | Acces siblty | Neat access | Appea-rance | Moder nness | Ele-gance | Envi-rnmnt |
| bi | me | 44 | bi | vr sm | bi | me | qr sm | si bi |
| bi | bi | 0 | ve bi | me | ve bi | ve bi | ve bi | ve bi |
| bi | bi | 29 | ve sm | bi | si sm | sm | ml sm | bi |
| bi | bi | 4 | bi | me | bi | me | me | ve bi |

The decision is based on the following characteristics:

1. Economical characteristics (price, reconstruction cost).
2. Technical characteristics (No. of rooms, No of bathrooms, No of toilets, quality of heating).
3. Sizes (size of house and garden, size of additional land).
4. Other characteristics (quality of garage, quality of cellar).
5. Infrastructure (distance from center, accessibility, neatness of access).
6. Aesthetical characteristics (global appearance, modernness, elegance).
7. Quality of environment.

The modeled decision situation is a multicriteria decision-making problem in which alternatives are characterized on the basis of the above listed characteristics. Note that among them, also non-measurable characteristics such as appearance, modernness and others are included. These are evaluated in the standard spaceless context $w = \langle 0, 0.4, 1\rangle$.

Values of these characteristics can be specified either directly, or using evaluative expressions, for example nice, ugly, very modern, etc. The evaluation could also be derived on the basis of special linguistic description, which may include many criteria having objective as well as subjective character. To give the reader a clear idea about the data, we summarized the data of four selected houses in table 1.

The hierarchy of linguistic descriptions is presented in Figure 4. As can be seen from the figure, some criteria are purely numerical, for example price, numbers of rooms, distance to the center, etc. and the other ones are qualitative. Among them are, for example, criteria such as "elegance", which is supposed to be

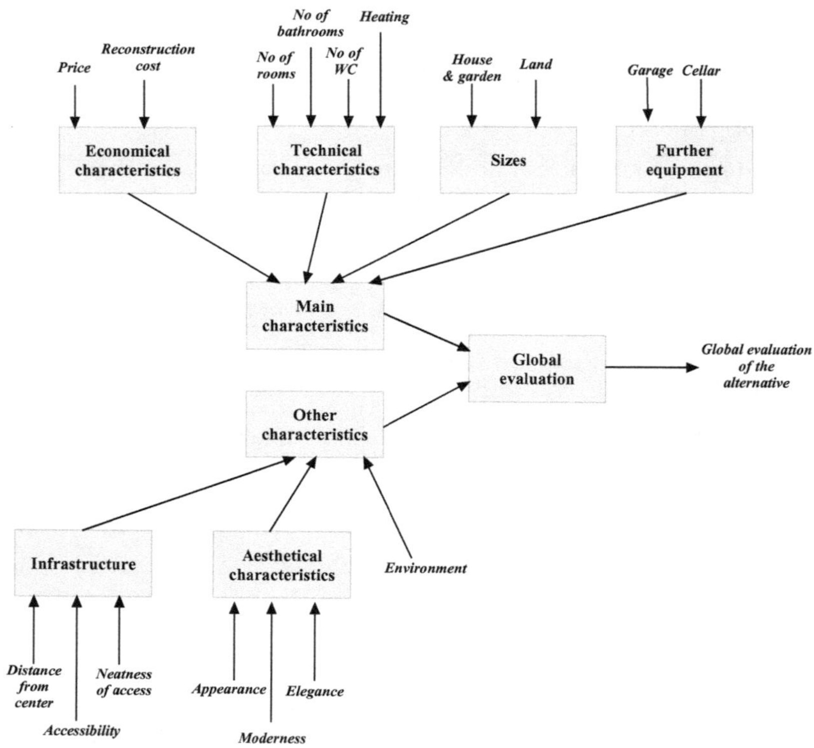


**Fig. 4.** Decision-making hierarchy for choosing the best house

evaluated by the expert using words such as *low, high, very high, medium*, etc., but also criteria such as "garage" or "heating", which includes not only its presence but also evaluated quality, state, or effectiveness (again using evaluative expressions).

Example of the linguistic description *Economical characteristics* is the following:

| Price | Reconstruction cost | | Economical characteristics |
|---|---|---|---|
| ve sm | ve sm | ⇒ | si bi |
| sm | sm | ⇒ | bi |
| sm | bi | ⇒ | me |
| me | me | ⇒ | ml sm |
| ml bi | ml bi | ⇒ | ve sm |
| bi | ve sm | ⇒ | ro bi |
| ve bi | no sm | ⇒ | ve sm |
| sm | me | ⇒ | ml bi |
| me | sm | ⇒ | ve bi |

(the rules are formed using evaluative expressions coded by the same shorts as in in table 1). In a similar way, we defined also the other linguistic descriptions according to Figure 4. The following is a linguistic description for the global evaluation.

| Main characteristics | Other characteristics | | Global evaluation |
|---|---|---|---|
| bi | bi | ⇒ | bi |
| ve bi | ve bi | ⇒ | si bi |
| si bi | ve bi | ⇒ | ex bi |
| bi | sm or me | ⇒ | ra bi |
| sm | sm | ⇒ | ve sm |
| sm | sm | ⇒ | ve bi |
| sm | bi | ⇒ | ml sm |
| me | me or bi | ⇒ | ro bi |
| sm | me | ⇒ | sm |
| me | qr sm | ⇒ | qr sm |

Let us remark that rules leading to refinement of the decision can still be added.

The results of the decision are summarized in the following table.

| House No. | Main characteristics | Other characteristics | Global evaluation numerical | Global evaluation linguistic |
|---|---|---|---|---|
| **1.** | **0.12** | **0.43** | **0.12** | **sm** |
| 2. | 0.41 | 0.26 | 0.26 | qr sm |
| 3. | 0.69 | 0.22 | 0.66 | ro bi |
| 4. | 0.12 | 0.79 | 0.19 | ml sm |
| 5. | 0.83 | 0.35 | 0.65 | ro bi |
| **6.** | **0.71** | **0.76** | **0.81** | **bi** |
| 7. | 0.65 | 0.88 | 0.65 | ro bi |
| 8. | 0.47 | 0.92 | 0.54 | vr bi |
| 9. | 0.46 | 0.50 | 0.44 | ml me |
| 10. | 0.41 | 0.88 | 0.69 | ro bi |
| 11. | 0.80 | 0.41 | 0.76 | ra bi |
| *12.* | *0.34* | *0.54* | *0.46* | *ml me* |
| 13. | 0.64 | 0.92 | 0.64 | ro bi |
| 14. | 0.52 | 0.58 | 0.52 | vr bi |
| 15. | 0.49 | 0.73 | 0.49 | vr bi |
| 16. | 0.53 | 0.84 | 0.53 | vr bi |
| 17. | 0.51 | 0.81 | 0.51 | vr bi |
| 18. | 0.05 | 0.88 | 0.15 | ra sm |
| **19.** | **0.73** | **0.74** | **0.82** | **bi** |
| 20. | 0.29 | 0.83 | 0.29 | vr sm |

The worst house is No. 1, the best is No. 19. However, the house No. 6 is practically the same. The house No. 12 is typically medium alternative. The reason why No. 1 is the worst is especially because of big reconstruction cost, small number of rooms, bathrooms and WC, and also infrastructure.

One can see that different influence of various criteria can be rendered using the evaluative expressions in a way that is well understandable to people. We can make tiny but at the same time sufficiently robust variations that lead to differences important for the concrete decision-maker.

### 3.3 Strategic Analysis and Evaluation of Time Series

When realizing their decisions, managers often rely on the data that have the form of times series. The task is then to evaluate them from the point of view of its historical and, possibly, also future development. The above described soft computing techniques can be effectively used for this purpose. Namely, the given time series can be analyzed using fuzzy transform and then its course can be evaluated using the PbLD technique. We will describe the method below.

A time series is a function

$$\{X_t \mid t = 1, \dots, T\} \subset \mathbb{R}, \qquad T \geq 3 \tag{16}$$

where $T \in \mathbb{N}$ a length. The time series can be decomposed into several components among which the most essential for the decision making is its trend. Therefore, we will consider decomposition of the times series into

$$X_t = D_t + R_t$$

where $D_t$, $t \in T$, is a trend and $R_t$ is a remainder which may include various kinds of influences having more or less random character (for example, seasonal influence, sickness of employees, failure of the equipment, etc.).

To estimate the the trend $D_t$, we will use the F-transform technique. First, we must specify the basic functions $A_1, \dots, A_{n_D}$ forming fuzzy partition of the interval $[1, T]$. The result are F-transform components

$$[F_1, \dots, F_{n_D}]. \tag{17}$$

Using the inverse F-transform (15), we obtain the following estimation of the trend:

$$\{D_t = X_{F,n_D}(t) \mid t = 1, \dots, T\}. \tag{18}$$

where $X_{F,n_D}(t)$ is an inverse transform of $X_t$ at time $t$.

Further step is forecasting the trend $D_t$. A simple moving average technique can be applied. According to our experiences, the length of moving averages should be 3–4. Thus, by moving averages we obtain new components $F_{n_D+1}$ (and possibly also $F_{n_D+2}$) and using the inverse F-transform compute the trend including its prediction using the formula

$$D_{T+k} = \sum_{i=n_D}^{n_D+2} F_i A_i(T+k)$$

where $A_i$ is the corresponding basic function.

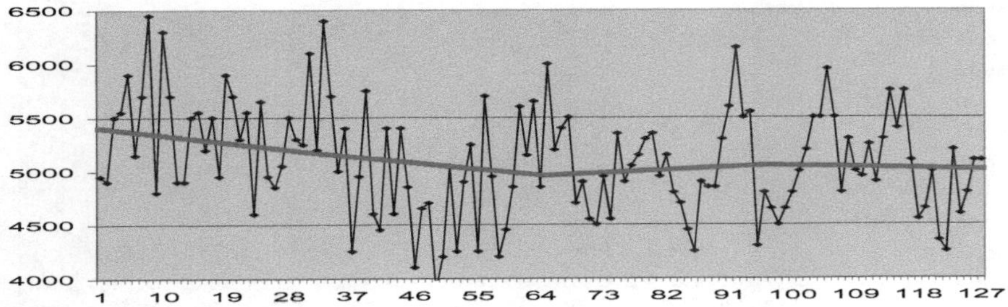


**Fig. 5.** Example of time series and its trend

To evaluate the trend, we will apply the theory of evaluative linguistic expressions. Let the F-transform components be given by (17). We will first normalize them to the range $[0, 1]$ simply by dividing each component by $\max\{F_1, \ldots, F_{n_D}\}$. The result are the normalized components

$$[F'_1, \ldots, F'_{n_D}]. \tag{19}$$

Then we compute the differences

$$\Delta F'_t = F'_t - F'_{t-1}, \qquad t = 2, \ldots n_D. \tag{20}$$

Each difference (20) expresses a partial tendency of the time series in the period determined by the length $h$ defined in Subsection 2.4 (this is half of the width of the corresponding basic function). The tendency can be evaluated using the evaluative linguistic expression.

For illustration, let us consider two times series in Figures 5 and 6. Their trend has been computed using F-transform with $h$ corresponding to 32 time points. Hence, the trend has four parts. On the basis of the differences (20) and using the theory of evaluative expressions we obtained evaluation of the trend in each part. The evaluation is, in principle, the local perception (8) derived on the basis of the context $[0, 0.08, 0.2]$. This context was applied separately for negative and positive value of (20) (note that this context means that maximal change corresponds to 20% of the value of the given component).

1. Time series from Figure 5: The local perceptions are "*−ro sm in part 1*", "*−ml sm in part 2*", "*sm in part 3*", and "*−si sm in part 4*". This can be interpreted as follows: the time series decreases in time period 1–64, increases in time period 65–96 and then again decreases. The decrease is first roughly small, then more or less small and finally only significantly small while the increase in the third time period is small.
2. Time series from Figure 6: The local perceptions are "*−ex bi in part 1*", "*−ml me in part 2*", "*−ve sm in part 3*", *and* "*sm in part 4*". Thus, the time series decreases in the time period 1 – 96 and increases a little in the last period. The decrease is first extremely big and then is slower while it is very small in the third period.

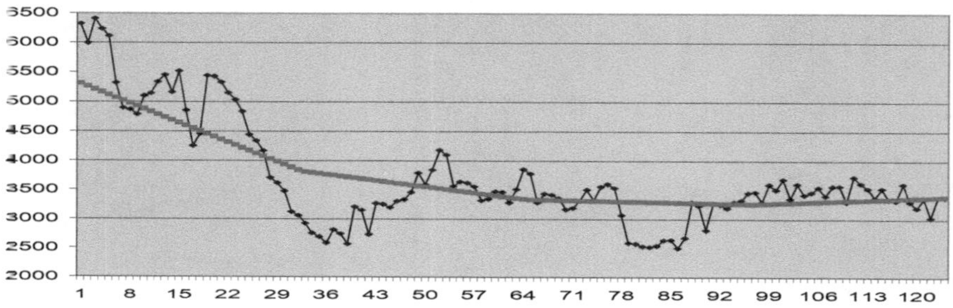


**Fig. 6.** Example of time series and its trend

The evaluation described above can be, of course, easily verified when watching the graphs. Note however, that our method can be automatized and applied to large number of time series in one moment. Moreover, the tendency need not always be immediately visible.

The information contained in the trend can be also used for evaluation of the global situation depending on the tendency of the given time series. For example, if the considered time series is data about gain of the company then such an evaluation can provide information how well is the company doing.

The evaluation can be done using PbLD on the basis of linguistic description having the following general form:

$$\mathscr{R}_1 := \text{IF } \Delta F'_{t-p} \text{ is } \mathscr{A}_{11} \text{ AND } \ldots \text{ AND } \Delta F'_t \text{ is } \mathscr{A}_{1p} \text{ THEN } V \text{ is } \mathscr{B}_1$$

$$\ldots\ldots\ldots\ldots\ldots\ldots\ldots\ldots\ldots\ldots\ldots\ldots\ldots\ldots\ldots\ldots\ldots\ldots$$

$$\mathscr{R}_m := \text{IF } \Delta F'_{t-p} \text{ is } \mathscr{A}_{m1} \text{ AND } \ldots \text{ AND } \Delta F'_t \text{ is } \mathscr{A}_{mp} \text{ THEN } V \text{ is } \mathscr{B}_m$$

where $p$ is the number of evaluated metaperiods the given time series and $\mathscr{A}_{ji}$, $\mathscr{B}_j$, $j = 1, \ldots, m, i = 1, \ldots, p$ are evaluative linguistic expressions. The variable $V$ is spaceless characteristics taking values $V$ in the context $\langle 0, 0.4, 1\rangle$ which expresses global characterization of the trend. If $V = 1$ then the trend is extremely good and if $V = 0$ then it is extremely bad.

The above mentioned methods can applied also for a more sophisticated time series analysis including their prediction. The latter information can be very important for managers especially when strategic decision is needed. More details about analysis of time series can be found in [11, 15].

## 4 Conclusion

In this paper, we introduced a general methodology which combines several intelligent soft-computing techniques to obtain relevant decision. The input information can be both in numerical as well as in verbal form. The used techniques include the theory of evaluative linguistic expressions which enables to assign linguistic expressions to numbers or vice-versa. This possibility is then effectively used when characterizing the behavior of time series. All such information can be combined into linguistic descriptions which are sets of conditional linguistic clauses call fuzzy/linguistic IF-THEN rules. The method which makes it possible to derive a conclusion on the basis of linguistic descriptions is called perception-based logical deduction. Using all this machinery, we can obtain a decision which well reflects the specific requirements of the decision-maker. We are convinced that the methodology is well suitable for managerial decision-making where combination of various kinds of information is necessary.

## References

[1] Bellman, R., Zadeh, L.: Decision making in a fuzzy environment. Management Science 17, 140–164 (1970)

[2] Bělohlávek, R., Novák, V.: Learning rule base of the linguistic expert systems. Soft Computing 7, 79–88 (2002)

[3] Calvo, T., Mayor, G., Mesiar, R. (eds.): Aggregation operators. New trends and applications. Physica-Verlag, Heidelberg (2002)

[4] Dvořák, A., Habiballa, H., Novák, V., Pavliska, V.: The software package LFLC 2000 - its specificity, recent and perspective applications. Computers in Industry 51, 269–280 (2003)

[5] Fodor, J., Roubens, M.: Fuzzy Preference Modelling and Multicriteria Decision Support. Kluwer, Dordrecht (1994)

[6] Hajičová, E., Partee, B.H., Sgall, P.: Topic-Focus Articulation, Tripartite Structures, and Semantics Content. Kluwer, Dordrecht (1998)

[7] Jie, L., Guangquan, Z., Da, R.: Multi-objective Group Decision Making: Methods, Software and Applications With Fuzzy Set Techniques. Imperial College Press, London (2007)

[8] Novák, V.: Perception-based logical deduction. In: Reusch, B. (ed.) Computational Intelligence, Theory and Applications, pp. 237–250. Springer, Berlin (2005)

[9] Novák, V.: A comprehensive theory of trichotomous evaluative linguistic expressions. Fuzzy Sets and Systems 159(22), 2939–2969 (2008)

[10] Novák, V., Perfilieva, I.: On the semantics of perception-based fuzzy logic deduction. International Journal of Intelligent Systems 19, 1007–1031 (2004)

[11] Novák, V., Štěpnička, M., Perfilieva, I., Pavliska, V.: Analysis of periodical time series using soft computing methods. In: Ruan, D., Montero, J., Lu, J., Martínez, L., D'hondt, P., Kerre, E. (eds.) Computational Intelligence in Decision and Control, p. 5560. World Scientific, New Jersey (2008)

[12] Perfilieva, I.: Fuzzy transform: Application to reef growth problem. In: Demicco, R.B., Klir, G.J. (eds.) Fuzzy Logic in Geology, pp. 275–300. Academic Press, Amsterdam (2003)

[13] Perfilieva, I.: Fuzzy transforms: theory and applications. Fuzzy Sets and Systems 157, 993–1023 (2006)

[14] Perfilieva, I., Haldeeva, E.: Fuzzy transformation. In: Proc. of the Joint 9th IFSA World Congress and 20th NAFIPS Int'l Conf. CD-ROM, Vancouver, Canada (2001)

[15] Perfilieva, I., Novák, V., Pavliska, V., Dvořák, A., Štěpnička, M.: Analysis and prediction of time series using fuzzy transform. In: Proc. IEEE World Congress on Computational Intelligence, pp. 3875–3879. IEEE Computational Intelligence Society, Hong Kong (2008)

[16] Saaty, T.: Fundamentals of Decision Making and Priority Theory With the Analytic Hierarchy Process. RWS Publications, Pittsburgh (2000)
[17] Sgall, P., Hajičová, E., Panevová, J.: The Meaning of the Sentence in Its Syntactic and Pragmatic Aspects. D. Reidel, Dordrecht (1986)
[18] Slowinski, R. (ed.): Fuzzy sets in decision analysis, operations research and statistics. Handbook of Fuzzy Sets Series. Kluwer, Dordrecht (1998)
[19] Zadeh, L.A.: Outline of a new approach to the analysis of complex systems and decision processes. IEEE Trans. on Systems, Man, and Cybernetics SMC-3, 28–44 (1973)
[20] Zadeh, L.A.: The concept of a linguistic variable and its application to approximate reasoning I, II, III. Information Sciences 8-9, 199–257, 301–357, 43–80 (1975)
[21] Zadeh, L.A.: Precisiated natural language. AI Magazine 25, 74–91 (2004)

8

# Supporting Decision Making via Verbalization of Data Analysis Results Using Linguistic Data Summaries

Janusz Kacprzyk and Sławomir Zadrożny

Systems Research Institute, Polish Academy of Sciences
ul. Newelska 6, 01-447 Warsaw, Poland
{kacprzyk,zadrozny}@ibspan.waw.pl

**Abstract.** We present how the conceptually and numerically simple concept of a fuzzy linguistic database summary can be a very powerful tool for gaining much insight into the essence of data that may be relevant for a business activity. The use of linguistic summaries provides tools for the verbalization of data analysis (mining) results which, in addition to the more commonly used visualization e.g. via a GUI, graphical user interface, can contribute to an increased human consistency and ease of use. The results (knowledge) derived are in a simple, easily comprehensible linguistic form which can be effectively and efficiently employed for supporting decision makers via the data driven decision support system paradigm. Two new relevant aspects of the analysis are also outlined which was first initiated by the authors. First, following Kacprzyk and Zadrożny (2009a) comments are given on an extremely relevant aspect of scalability of linguistic summarization of data, using their new concept of a conceptual scalability that is crucial for large applications. Second, following Kacprzyk and Zadrożny (2009b) it is further considered how linguistic data summarization is closely related to some types of solutions used in natural language generation (NLG), which can make it possible to use more and more effective and efficient tools and techniques developed in this another rapidly developing area. An application for a computer retailer is outlined.

## 1 Introduction

Decision making has always been a primordial concern of human being, at all levels of: an individual, a group of individuals, institutions, organizations, even nations and global groups of nations. Quite naturally, people have been trying since the very beginning of human race to first understand intricacies of the decision making process, reasons for success or failure, and finally to develop decision making procedure that could have been better that what people have been accustomed to. Needless to say that the complexity of decision making processes has been increasing over the centuries reaching finally in recent decades or years the level of unprecedented difficulty related to so many different actors (agents), criteria, points of view, etc. which have to be taken into account. The world wide economic and financial crisis is just a reflection of those inherent difficulties.

As it is the case of all problems of crucial importance, sciences has been trying to conceptualize and analyze decision making processes for many decades hoping

E. Rakus-Andersson et al. (Eds.): Recent Advances in Decision Making, SCI 222, pp. 121–143.
springerlink.com 

to find tools and techniques that would be helpful to find better decisions. The involvement of science has been particularly strong in a couple of recent decades. In general, researchers have been trying to find tools and techniques to be able to tackle complex, nontrivial decision making problem occurring in real life, exemplified by strategic planning, environmental pollution control, etc. All those problems involve a myriad of different, often contradicting criteria, multiple players (decision makers, agents, actors,…) with contradicting value systems, preferences, access to information, etc., a complex dynamics, often highly nonlinear resulting in chaotic or emergent behavior, etc. For our purposes, a clear indication that could have been seen from those studies was that to effectively and efficiently solve complex real world problems, human decision makers should be supported by computerized decision support systems (DSSs), but not replaced by those systems!

This is also philosophy of our approach presented in this paper. We will show some solution that should be proper for the class of problems considered. However, before coming to what we will mean by a decision support system, which kind of such systems may be the most appropriate, and which additional tools should be provided to amplify these capabilities, we should start with some more general remarks on decision making.

Research in decision making has been concentrated for a long time on the development of mathematical models that would try to describe the situation under consideration (preferences, mathematical models, performance functions, solution concepts, etc.). Numerous models have been proposed, both descriptive and prescriptive, involving single and multiple criteria and decision makers, dynamics, etc. All this has been done in a strict mathematical setting, notably that of utility maximization.

Modern approaches to real world decision making propose a considerable departure from that paradigm (cf. Wierzbicki, Makowski and Wessels, 2000) by basically speaking about *good decisions* (not *optimal* as in most traditional approaches) but above all stating that a *decision making process* is concerned. Basically, a decision making process involves more elements and aspects that traditional decision making models did, notably:

- Use of own and external knowledge,
- Involvement of various „actors", aspects, etc.
- Individual habitual domains,
- Non-trivial rationality,
- Different paradigms, when appropriate.

A good example of such a decision making process is Peter Checkland's (1989, 1999) so-called *deliberative decision making* (which is an important element of his *soft approach to systems analysis*). The essence of deliberative (soft) decision making may be subsumed as follows: to solve a complex real world decision making problem we should:

- Perceive the whole picture,
- Observe it from all angles (actors, criteria,...)
- Find a *good* decision using *knowledge* and *intuition*.

Further, it is emphasized in modern approaches that the decision making process involves:

- Recognition,
- Deliberation and analysis,
- Gestation and enlightenment (the so-called „eureka!", „aha" effects),
- Rationalization,
- Implementation.

and:

- Is heavily based on data, information and knowledge, and human specific characteristics (intuition, attitude, natural language for communication and articulation,...),
- Needs number crunching, but also more „delicate" and sophisticated „intelligent" analyses,
- Heavily relies on computer systems, and is able to employ and exploit a synergistic human-computer interaction, notably using tools and techniques that are more adequate to human cognitive capabilities, notably using graphical displays, i.e. *visualization*, and (quasi)natural language, i.e. *verbalization* during the problem formulation, solution, displaying of results, etc.

A natural consequence is that to effectively and efficiently solve modern real world decision making problems, decision makers should be supported by some computerized systems, called *decision support systems* (DSSs) – see, e.g., Alter (1990), Holsapple and Whinston (1996), Sprague and Watson (1996), etc. The essence of the problems implies that in the development of such systems emphasis should be on:

- Ill/semi/un-structured questions and problems,
- Non-routine, one of a kind answers,
- A flexible combination of analytical models and data,
- Various kinds of data, e.g. numeric, textual, verbal,...
- Interactive interface (e.g. GUI, LUI),
- Iterative operation („what if"),
- Support of various decision making styles,
- Support of alternate decision making passes, etc.

All the above mentioned phases are based on data, information and knowledge, meant here as:

- Data - raw facts;
- Information - data in a context relevant to an individual, team or organization,
- Knowledge - an individual's utilization of information and data complemented by an unarticulated expertise, skills, competencies, intuitions, experience and motivations (for simplicity, we will not go further to include wisdom as the highest level).

As to knowledge, which is the most relevant (cf. Holsapple and Whinston, 1996), it can be:

- Explicit, expressed in words or numbers, and shared as data, equations, specifications, documents, and reports; can be transmitted between individuals and formally recorded,

- Tacit, highly personal, hard to formalize, and difficult to communicate or share with others; technical (skills or crafts), and cognitive (perceptions, values, beliefs, and mental models).

and both types are relevant for decision making processes and hence for the DSSs.

Historically, DSSs practically appeared in the mid-1960s with the development of IBM 360 and a wider use of distributed, time-sharing computing, and have been since that time a topic of intensive research and development.

One can distinguish the following basic types of DSSs (cf. the famous Dan Power's classification: www.dssresources.com):

- Data driven,
- Communication driven and group DSSs,
- Document driven,
- Model driven,
- Knowledge driven,
- Web based and interorganizational.

Roughly speaking:

- Data Driven DSSs emphasize access to and manipulation of internal company data and sometimes external data, and may be based –from the low to high level – first on simple file systems with query and retrieval tools, then data warehouses, and finally with On-line Analytical Processing (OLAP) or data mining tools.
- Communications Driven DSSs use network and communications technologies to facilitate collaboration and communication.
- Group GDSSs are interactive, computer-based systems that facilitate solution of unstructured problems by a set of decision-makers working together as a group.
- Document Driven DSSs integrate a variety of storage and processing technologies for a complete document retrieval and analysis; documents may contain numbers, text, multimedia.
- Model Driven DSSs emphasize access to and manipulation of a model, e.g., statistical, financial, optimization and/or simulation; use data and parameters, but are not usually data intensive.
- Knowledge Driven DSSs are interactive systems with specialized problem-solving expertise consisting of knowledge about a particular domain, understanding of problems within that domain, and "skill" at solving some of these problems.
- Web based DSSs are computerized system that deliver decision support related information and/or tools to a manager/analyst using a "thin-client" Web browser (e.g. Microsft Internet Explorer®); TCP/IP protocol, etc.

Notice that though the model driven DSSs may seem the most natural and developed, in many situations the development of an adequate, nontrivial model may be very difficult or too costly, and in many cases mathematical models may be simply unknown. On the other hand, since the DSSs are meant to support human decision makers, and not to replace them, maybe some non-model driven DSSs can be very useful and can do the job.

In this paper we concentrate on the data driven DSSs, and in particular show how the use of Zadeh's computing with words and perception paradigm (cf. Zadeh and Kacprzyk, 1999a, b) through fuzzy linguistic database summaries, and indirectly fuzzy querying, can open new vistas in data driven DSSs (and also, to some extent, in knowledge driven and Web based DSSs). We will argue for the simplicity of this approach and its high scalability mainly due to an extensive use of natural language which is the only fully natural means of articulation and communication by the humans.

Basically, the role of a data driven DSS is to help decision makers make rational use of (vast) amounts of data that exist in their environment (e.g. a company or institution) within which they operate. From those data some useful, relevant, nontrivial dependencies should be found, and their discovery is not trivial and requires the use of some sophisticated IT tools, notably those of intelligent systems.

One of interesting and promising approaches meant for these purposes is to derive linguistic summaries of a set of data (database). Here we discuss linguistic summarization of data sets in the sense of Yager (1982, 1989 - 1996) (for some extensions and other related issues, see, e.g., Kacprzyk and Yager (2001), Kacprzyk, Yager and Zadrożny (2000), Rasmussen and Yager (1996 – 1999), Yager and Rubinson (1981), etc.).

In this approach linguistic summaries are derived as linguistically quantified propositions, exemplified – when the data in question concern employees – by "most of the employees are young and well paid", with which a degree of validity is associated. Basically, in the source Yager's (1982) work that degree of validity was meant to be the degree of truth of a linguistically quantified proposition that constitutes a summary. This was shown to be not enough, and other validity (quality) indicators were proposed, also in the above Yager's works. As a relevant further attempt, we can mention George and Srikanth's (1996) solution in which a compromise between the specificity and generality of a summary is sought, and then some extension in which a weighted sum of 5 quality indicators is employed as given in Kacprzyk and Yager (2001) and Kacprzyk, Yager and Zadrożny (2000).

In this paper we also follow the philosophy of Kacprzyk and Zadrożny's (1998, 1999, 2000a, b, c), Kacprzyk's (1999), and Zadrożny and Kacprzyk's (1999) idea of an interactive approach to linguistic summaries. Basically, since a fully automatic generation of linguistic summaries is not feasible at present, an interaction with the user is assumed for the determination of a class of summaries of interest. This is done via Kacprzyk and Zadrożny's (1994 - 1996) fuzzy querying add-on to Microsoft Access.

Then, following and extending Kacprzyk and Zadrożny (2002, 2005a), we show that by relating various types of linguistic summaries to fuzzy queries, with various known and sought elements, we can arrive at a hierarchy of prototypical forms, or – in Zadeh's (2002, 2006) terminology – *protoforms*, of linguistic data summaries. We advocate that they are a very powerful conceptual idea because they provide a simple structural expression, with a comprehensible semantics, of even the most complicated linguistic summaries.

Notice that, first, through the use of natural language to present (verbalize) the essence of data with respect to an aspect in question we certainly attain a high, maybe even ultimate human consistency because natural language is the only fully natural means of articulation and communication of a human being. Moreover, through natural language we attain an ultimate scalability as natural language can express in a human comprehensive way information no matter how large the data set is. Simple linguistically quantified propositions with which data summaries are equated may semantically be adequate as representations of data sets of any size. Protoforms of linguistic summaries provide a uniform, easily comprehensible form of linguistic summaries for any size of data sets, and virtually all intentions and information needs of the user too. And last but not least, natural language summaries are comprehensible to individuals, small and larger groups, people from different backgrounds, people coming from various geographic locations, sexes, age groups, etc. These issues of scalability of linguistic data summaries have been dealt with in detail in Kacprzyk and Zadrożny (2009a) in which the concept of a *conceptual scalability* has been introduced as a complement to the technical scalability normally considered.

Another important aspect is whether linguistic data summaries are related to some other well established techniques, and in this respect Kacprzyk and Zadrożny (2009b) have indicated that they directly correspond to some specific, so-called template based, techniques of natural language generation (cf. Reiter and Dale, 2000), but extend those traditional techniques by making it possible to account for the inherent imprecision of natural language.

We relate the process of derivation of linguistic summaries more explicitly to the construction and implementation of a data driven DSS. We also present an implementation for a sales database of a computer retailer. We show that the linguistic summaries obtained may be very useful for supporting decision making.

## 2 Linguistic Data Summaries: The Concept and Extensions

In this paper we consider first a simple yet effective and efficient approach to the linguistic summarization of data sets proposed by Yager (1982), and then presented in a more advanced, and implementable form by Kacprzyk and Yager (2001), and Kacprzyk, Yager and Zadrożny (2000). We have:

- $V$ is a quality (attribute) of interest, e.g. salary in a database of workers,
- $Y = \{y_1, \ldots, y_n\}$ is a set of objects (records) that manifest quality $V$, e.g. the set of workers; hence $V(y_i)$ are values of quality $V$ for object $y_i \in Y$;
- $D = \{V(y_1), \ldots, V(y_n)\}$ is a set of data (the "database" on question)

A *linguistic summary* of a data set consists of:

- a summarizer $S$ (e.g. young),
- a quantity in agreement $Q$ (e.g. most),
- truth $T$ - e.g. 0.7,

as, e.g., "$T$(*most* of employees are *young*)=0.7". The truth $T$ may be meant in a more general sense, e.g. as validity or, even more generally, as some quality or goodness of a linguistic summary.

Basically, for a set of data $D$, we can hypothesize any appropriate summarizer $S$ and any quantity in agreement $Q$, and the assumed measure of truth will indicate the truth of the statement that $Q$ data items satisfy the statement (summarizer) $S$.

Notice that we consider here some specific, basic form of a linguistic summary. Notice also that we discuss here the linguistic summarization of sets of numeric values only. One can clearly imagine the linguistic summarization of both symbolic values or textual information but this is beyond the scope of this paper. We do not consider here neither some other approaches to the linguistic summarization of data sets that are based on a different philosophy, cf. Bosc et al. (2002), Dubois and Prade (2000), Raschia and Mouaddib (2002) or Rassmussen and Yager (1996,1997a,1997b,1999).

As to the forms of the particular elements of a linguistic summary considered, the summarizer $S$ is a linguistic expression semantically represented by a fuzzy set like, for instance "young" would be represented as a fuzzy set in the universe of discourse as, e.g., $\{1, 2, ..., 90\}$, i.e. containing possible values of the human age, and "young" could be given as, e.g., a fuzzy set with a non-increasing membership function in that universe such that, in a simple case of a piecewise linear membership function, the age up to 35 years is for sure "young", i.e. the grade of membership is equal to 1, the age over 50 years is for sure "not young", i.e. the grade of membership is equal to 0, and for the ages between 35 and 50 years the grades of membership are between 1 and 0, the higher the age the lower its corresponding grade of membership. This kind of a summarizer exemplified by "young" can clearly be extended to, e.g, "*young* and *well paid*".

In reality, the most interesting are more sophisticated, *human-consistent* summarizers (concepts) as, e.g.:

- productive workers,
- stimulating work environment,
- difficult orders, etc.

defined by a complicated *combinations of attributes*, e.g.: a hierarchy (not all attributes are of the same importance), the attribute values are ANDed and/or ORed, $k$ out of $n$, *most*, etc. of them should be accounted for, etc. Such summarizers need some specific tools and techniques.

The quantity in agreement, $Q$, is an indication of the range of data satisfying the summarizer. We use also here linguistic terms represented by fuzzy sets. Basically, two types of such a linguistic quantity in agreement can be used:

- absolute as, e.g., "about 5", "more or less 100", "several", and
- relative as, e.g., "a few", "more or less a half", "most", almost all"etc.

and both are the so-called fuzzy linguistic quantifiers (cf. (Zadeh, 1983)) that can be handled by fuzzy logic.

The calculation of the truth (or, more generally, validity) of the linguistic summary considered is equivalent to the calculation of the truth value of a corresponding linguistically quantified statement (e.g., "*most* of the employees are *young*"). This can be calculated by using two most relevant techniques: Zadeh's (1983) calculus of linguistically quantified statements (cf. (Zadeh and Kacprzyk, 1999ab)) or Yager's

(1988) OWA operators (cf. (Yager and Kacprzyk, 1997)). In what follows we briefly remind the basics of these two techniques.

A linguistically quantified proposition, e.g., "most experts are convinced", is written as "$Qy's$ are $F$", where $Q$ is a linguistic quantifier (e.g., most) $Y = \{y\}$ is a set of objects (e.g., experts), and $F$ is a property (e.g., convinced). Importance $B$ may be added yielding "$QBy's$ are $F$", e.g., "most ($Q$) of the important ($B$) experts ($y$'s) are convinced ($F$)". The problem is to find $\text{truth}(Qy's \text{ are } F)$ or $\text{truth}(QBy's \text{ are } F)$, respectively, knowing $\text{truth}(y \text{ is } F), \forall y \in Y$ which is done using Zadeh's (1983) fuzzy logic based calculus of linguistically quantified propositions as follows.

Property $F$ and importance $B$ are fuzzy sets in $Y$, and a (proportional, nondecreasing) linguistic quantifier $Q$ is assumed to be a fuzzy set in [0,1] as, e.g.

$$\mu_Q(x) = \begin{cases} 1 & \text{for } x \geq 0.8 \\ 2x - 0.6 & \text{for } 0.3 < x < 0.8 \\ 0 & \text{for } x \leq 0.3 \end{cases} \tag{1}$$

Then, due to Zadeh (1983)

$$\text{truth}(Qy's \text{ are } F) = \mu_Q[\tfrac{1}{n} \textstyle\sum_{i=1}^{n} \mu_F(y_i)] \tag{2}$$

$$\text{truth}(QBy's \text{ are } F) = \mu_Q[\textstyle\sum_{i=1}^{n} (\mu_B(y_i) \wedge \mu_F(y_i)) / \sum_{i=1}^{n} \mu_B(y_i)] \tag{3}$$

An OWA operator (Yager, 1988; Yager and Kacprzyk, 1997) of dimension $p$ is a mapping $O : [0,1]^p \to [0,1]$ if associated with $O$ is a weighting vector, $W = [w_1, \ldots, w_p]^T$, $w_i \in [0,1], w_1 + \cdots + w_p = 1$, and

$$O(x_1, \ldots, x_p) = w_1 b_1 + \cdots w_p b_p = W^T B \tag{4}$$

where $b_i$ is the $i$-th largest element among $x_1, \ldots, x_p$, $B = [b_1, \ldots, b_p]$.

The OWA weights may be found from the membership function of $Q$ due to (cf. Yager, 1988):

$$w_i = \mu_Q\left(\frac{i}{p}\right) - \mu_Q\left(\frac{i-1}{p}\right) \quad \text{for } i = 1, \ldots, p \tag{5}$$

The OWA operators can model a wide array of aggregation operators (including linguistic quantifiers), from $w_1 = \ldots = w_{p-1} = 0$ and $w_p = 1$ which corresponds to "all", to $w_1 = 1$ and $w_2 = \ldots = w_p = 0$ which corresponds to " at least one", through all intermediate situations, and that is why they are widely employed.

An important case is when with the OWA operator importance qualification of the particular pieces of data is associated. Suppose that with the data $A = [a_1, \ldots, a_p]$, a vector of importances $V = [v_1, \ldots, v_p]$, such that $v_i \in [0,1]$ is the importance of

$a_i, i = 1, \ldots, p$, $v_1 + \cdots v_p = 1$, is associated. Then, for an *ordered weighted averaging operator with importance qualification*, denoted $O_V$, Yager (1988) proposed that, first, some redefinition of the OWA's weights $w_i$'s into $\overline{w}_i$'s is performed, and (4) becomes

$$O_V(x_1, \ldots, x_p) = \overline{w}_1 b_1 + \cdots \overline{w}_p b_p = \overline{W}^T B \tag{6}$$

where

$$\overline{w}_j = \mu_Q\left(\frac{\sum_{k=1}^{j} u_k}{\sum_{k=1}^{p} u_k}\right) - \mu_Q\left(\frac{\sum_{k=1}^{j-1} u_k}{\sum_{k=1}^{p} u_k}\right) \tag{7}$$

where $u_k$ is the importance of $b_k$, i.e. the $k$-largest element of $A$. This concludes our brief reminder of the basics of the Zadeh's calculus of linguistically quantified propositions and Yager's OWA operators.

The basic validity criterion, i.e. the truth of a linguistically quantified statement given by (2) and (3), is certainly the most natural and important but it does not grasp all aspects of a linguistic summary. Some other, additional quality criteria have been proposed in the literature, starting from some measure of informativeness in the source Yager's (1982) paper, through some measures given by George and Srikanth (1996), to a comprehensive set of measures given by Kacprzyk and Yager (2001), and Kacprzyk, Yager and Zadrożny (2000) who have proposed:

- a truth value (which basically corresponds to the degree of truth of a linguistically quantified proposition representing the summary given by, say, (2) or (3)),
- a degree of imprecision,
- a degree of covering,
- a degree of appropriateness,
- a length of a summary.

Unfortunately, due to lack of space, we will not discuss these measures referring the interested readers to the papers cited. The essence of these measures can be summarized as follows:

The **degree of truth**, $T_1$, is the basic validity criterion which results directly from the use of Zadeh's (1983, 1985) calculus of linguistically quantified propositions.

The **degree of imprecision** is an obvious and important validity criterion. Basically, a linguistic summary (e.g. „On almost all winter days the temperature is rather cold") has a very high degree of truth yet it is not useful due to the very imprecise character of the summarizer „rather cold". Notice that the degree of imprecision depends on the form of the summary only and not on the data and its calculation does not require the searching of the database.

The **degree of covering** says how many objects in the data set corresponding to the query are covered by the particular summary. The value of this degree depends clearly on the contents of the database.

The **degree of appropriateness,** which is probably the most relevant measure, describes how characteristic for the particular database the summary found is.

The **length** of a summary is relevant because a long summary is not easily comprehensible by the human user.

Now, denoting the above degrees as $T_1$, $T_2$, $T_3$, $T_4$, $T_5$, with the respective weighs, $w_1$, $w_2$, $w_3$, $w_4$, $w_5$, assigned to the particular degrees of validity (with values from the unit interval, the higher, the more important such that $\sum_{i=1,2,...,5} w_i = 1$), the (total) degree of validity, $T$, of a particular linguistic summary is defined as the weighted average of the above 5 degrees of validity, i.e.:

$$T = T(T_1, T_2, T_3, T_4, T_5; w_1, w_2, w_3, w_4, w_5) = \sum_{i=1,2,...,5} w_i T_i \quad (8)$$

and the problem is to find an optimal summary, $S^* \in \{S\}$, such that

$$S^* = \arg\max_S \sum_{i=1,2,...,5} w_i T \quad (9)$$

The definition of weights, w1,...,w5, is a problem in itself, and will not be dealt with in more detail. The weights can be predefined or elicited from the user, e.g, using the Saaty's AHP technique (Saaty, 1980).

As we have already mentioned, the linguistic summarization meant as the solution of (9) may be numerically difficult, and some non-exhaustive search techniques, normally based on some heuristics, should be employed but this will not be considered here. Therefore, the linguistic summarization process is not well scalable in the traditional sense but, using the concept of cognitive (perceptual) scalability introduced by Kacprzyk and Zadrożny (2009a), it may be said totally conceptually (perceptually) scalable because it is comprehensible to a human being, either an individual or a group of individuals, no matter what size of the data set is. This is a direct result of, on the one hand, the use of natural language, which is the only fully natural means of articulation and communication of a human being, and – on the other hand – of a simple and intuitively appealing form of a linguistic summary which basically says what most of the data exhibit, i.e. what *usually happens* (holds).

A fully automatic determination of a best linguistic summary, i.e. the solution of (9) may be therefore infeasible in practice, and therefore Kacprzyk and Zadrożny (1998, 2001a) proposed an *interactive approach* with a *user assistance* in the definition of summarizers, by the indication of attributes and their combinations of interest. This proceeds via a user interface of a fuzzy querying add-on. Basically, the queries (referring to summarizers) allowed are:

- *simple* as, e.g., "salary is *high*"
- *compound* as, e.g., "salary is *low* AND age is *old*"
- *compound (with quantifier)*, as, e.g., "*most* of {salary is *high*, age is *young*, ..., training is *well above average*}.

In Kacprzyk and Zadrożny (1994, 1995a, 1995b, 2001b), a conventional DBMS is used, and a fuzzy querying tool is developed to allow for queries with fuzzy (linguistic) elements of the "simple", "compound" and "compound with quantifier" types. This fuzzy querying system (add-in) has been developed for Microsoft Access® but it is clearly applicable to any DBMS. The main problems to be solved are here: (1) how to extend the syntax and semantics of the query, and (2) how to provide an easy way of eliciting and manipulating those terms by the user. This will now be briefly described, emphasizing those aspects which are relevant.

FQUERY for Access is embedded in the native Microsoft Access's environment as an add-in. All the code and data is put into a database file, a *library*, installed by the user. Definitions of attributes, linguistic terms etc. are maintained in a dictionary (a set of regular tables), and a mechanism for putting them into the Query-By-Example (QBE) sheet (grid) of the Microsoft Access' interface is provided. Linguistic terms are represented within a query as parameters, and a query transformation is performed to provide for their proper interpretation during the query execution.

FQUERY for Access is an add-in that makes it possible to use linguistic (fuzzy) terms in queries:

- fuzzy values, exemplified by *low* in "profitability is *low*",
- fuzzy relations, exemplified by *much greater than* in "income is *much greater than* spending", and
- linguistic quantifiers, exemplified by *most* in "*most* conditions have to be met",

where the elements of the first two types are elementary building blocks of fuzzy queries in FQUERY for Access. They are meaningful in the context of numerical fields only.

If a field (or column, in the relational database parlance) is to be used in a query in connection with a fuzzy value, it has to be defined as an *attribute* by specifying two numbers: the attribute's values lower (LL) and upper (UL) limit. They set the interval which the field's values are assumed to belong to, according to the user. For example, for the *age* (of a person), the reasonable interval would be, e.g., [18, 65], in a particular context. Such a concept of an attribute makes it possible to universally define fuzzy values.

*Fuzzy values* are defined as fuzzy sets on [-10, +10]. Then, *the matching degree* $md(\cdot,\cdot)$ of a simple condition referring to attribute AT and fuzzy value FV against a record R is calculated by:

$$md(\mathrm{AT} = \mathrm{FV}, \mathrm{R}) = \mu_{\mathrm{FV}}(\tau(\mathrm{R}(\mathrm{AT}))) \tag{10}$$

where: R(AT) is the value of attribute AT in record R, $\mu_{\mathrm{FV}}$ is the membership function of fuzzy value FV, $\tau$: $[\mathrm{LL}_{\mathrm{AT}},\mathrm{UL}_{\mathrm{AT}}]\rightarrow[-10,10]$ is the mapping from the interval defining AT onto [-10,10] so that we may use the same fuzzy values for different fields. A meaningful interpretation is secured by $\tau$ which makes it possible to treat all fields domains as ranging over the unified interval [-10,10]. For simplicity, it is normally assumed, also here, that the membership functions of fuzzy values are trapezoidal as in Figure 1 and $\tau$ is assumed linear.

*Linguistic quantifiers* provide for a flexible aggregation of simple conditions. In FQUERY for Access the fuzzy linguistic quantifiers are defined in Zadeh's (1983, 1985) sense, as fuzzy sets on [0, 10] interval instead of the original [0, 1] – cf. *most* given as (1). They may be interpreted either using original Zadeh's (1983) approach or via the OWA operators, cf. Yager (1988) or Yager and Kacprzyk (1997)); Zadeh's interpretation will be considered in what follows. The membership functions of fuzzy linguistic quantifiers are assumed piece-wise linear, hence two numbers from [0, 10]

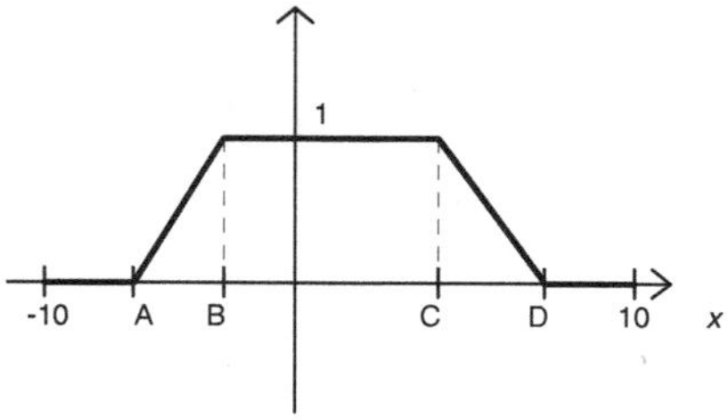


**Fig. 1.** An example of the membership function of a fuzzy value

are needed. Again, a mapping from [0, *N*], where *N* is the number of conditions aggregated, to [0, 10] is employed to calculate the matching degree of a query. More precisely, the matching degree, $md(\cdot,\cdot)$, for the *query* "*Q* of *N* conditions are satisfied" for record R is equal to

$$md(Q, \text{condition}_i, \text{R}) = \mu_{\text{Q}}[\tau(\sum_i md(\text{condition}_i, \text{R}))] \tag{11}$$

We can also assign different importance degrees for particular conditions. Then, the aggregation formula is equivalent to (3). The importance is identified with a fuzzy set on [0,1], and then treated as property *B* in (3).

FQUERY for Access has been designed so that queries containing fuzzy terms are still syntactically correct Access's queries. It has been attained through the use of parameters. Basically, Access represents the queries using SQL. Parameters, expressed as strings limited with brackets, make it possible to embed references to fuzzy terms in a query. We have assumed special naming convention for parameters corresponding to particular fuzzy terms. For example, a parameter like:

| | |
|---|---|
| [FfA_FV *fuzzy value name*] | will be interpreted as a fuzzy value |
| [FfA_FQ *fuzzy quantifier name*] | will be interpreted as a fuzzy quantifier |

Before a fuzzy term may be used in a query, it has to be defined using the toolbar provided by FQUERY for Access and stored internally. This feature, i.e. maintenance of dictionaries of fuzzy terms defined by users, strongly supports our approach to data summarization discussed in this chapter. In fact, the package comes with a set of predefined fuzzy terms but the user may enrich the dictionary too.

When the user initiates the execution of a query it is automatically transformed by appropriate FQUERY for Access's routines and then run as a native query of Access. The transformation consists primarily in the replacement of parameters referring to fuzzy terms by calls to functions implemented by the package which secure a proper interpretation of these fuzzy terms. Then, the query is run by Access as usually.

FQUERY for Access provides its own toolbar. There is one button for each fuzzy element, and the buttons for declaring attributes, starting the querying, closing the toolbar and for help (cf. Figure 2).

Details can be found in Kacprzyk and Zadrożny (1994 – 1995b).

Clearly, fuzzy queries directly correspond to summarizers in linguistic summaries which was first formally shown by Kacprzyk and Zadrożny (1998). Thus, the derivation of a linguistic summary may proceed in an interactive (user assisted) way as follows:

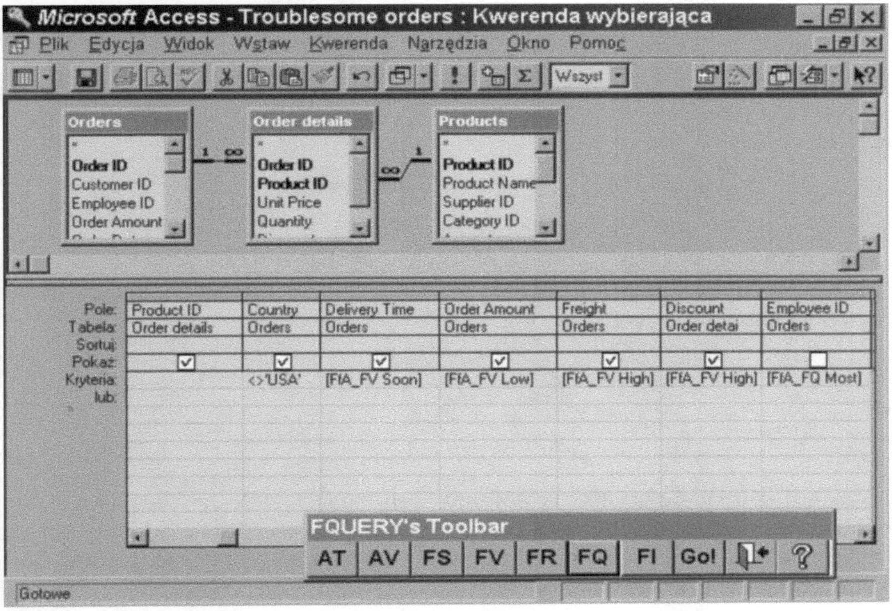


**Fig. 2.** Composition of a fuzzy query

- the user formulates a set of linguistic summaries of interest (relevance) using the fuzzy querying add in,
- the system retrieves records from the database and calculates the validity of each summary adopted, and
- some best (most appropriate) linguistic summary is chosen.

Therefore, we can restate the linguistic summarization in the fuzzy querying context. First, (2) may be interpreted as:

$$\text{"Most records match query } S\text{"} \tag{11}$$

where $S$ replaces $F$ in (2) since we refer here directly to the concept of a summarizer (of course, $S$ is in fact the whole condition, e.g., price = *high*, while $F$ is just the fuzzy value, i.e. *high* in this condition; this should not lead to confusion).

Similarly, (3) may be interpreted as:

$$\text{"Most records meeting conditions } B \text{ match query } S\text{"} \tag{12}$$

Thus, (12) says something only about a subset of records specified by (11). In database terminology, $B$ corresponds to a *filter* and (12) claims that *most* records passing through $B$ match query $S$. Moreover, since the filter may be fuzzy, a record may pass through it to a degree from [0,1].

Looking at (11) and (12), which specify the user's interest and intent as to linguistic data summaries put in the context of database querying, a very relevant idea was proposed by Kacprzyk and Zadrożny (2002, 2005a) that the concept of a *protoform* in the sense of Zadeh (2002, 2006) is highly relevant. A protoform is defined as an abstract prototype, that is, in our context, for the query (summary) given by (11) and (12) as follows, respectively:

$$\text{"}\textit{Most R's are S}\text{"} \tag{13}$$

and

$$\text{"}\textit{Most BR's} \text{ are } S\text{"} \tag{14}$$

where $R$ means "records", $B$ is a filter, and $S$ is a query.

Protoforms can obviously form a hierarchy, so that we can define higher level (more abstract) protoforms, for instance replacing *most* by a general linguistic quantifier $Q$, obtaining, respectively:

$$\text{"}QR\text{'s are } S\text{"} \tag{15}$$

and

$$\text{"}QBR\text{'s are } S\text{"} \tag{16}$$

Obviously, the more abstract protoforms correspond to cases in which we assume less about summaries sought, with two limit cases: (1) we assume a totally abstract protoform, and (2) we assume all elements of a protoform to be given. In case 1 data summarization will be extremely time consuming, as the search space may be enormous, but may produce interesting, unexpected views on data. In case 2 the user has to guess a good candidate formula for summarization but the evaluation is fairly simple, just equivalent to the answering of a (fuzzy) query. Thus, the second case refers to the summarization known as *ad hoc queries.*

Then, going further along this line, we can show in Table 1 a classification of linguistic summaries into 5 basic types corresponding to protoforms of a more and more abstracted form.

**Table 1.** Classification of linguistic summaries

| Type | Given | Sought | Remarks |
|---|---|---|---|
| 1 | $S$ | $Q$ | Simple summaries through ad-hoc queries |
| 2 | $S$ $B$ | $Q$ | Conditional summaries through ad-hoc queries |
| 3 | $Q$ $S^{structure}$ | $S^{value}$ | Simple value oriented summaries |
| 4 | $Q$ $S^{structure}$ $B$ | $S^{value}$ | Conditional value oriented summaries |
| 5 | nothing | $S$ $B$ $Q$ | General fuzzy rules |

where $S^{structure}$ denotes that attributes and their connection in a summary are known, while $S^{value}$ denotes a summarizer sought.

Type 1 summaries may be easily derived by a simple extension of fuzzy querying given by Kacprzyk and Zadrożny's (2001b) FQUERY for Access. The user has to construct a query, a candidate summary, and the derivation module has just to find the fraction of rows matching this query and a linguistic quantifier best denoting this fraction. A Type 2 summary is a straightforward extension of Type 1 by adding a fuzzy filter. Type 3 summaries require much more effort, and are concerned with the determination of typical (exceptional) values of an attribute. So, query $S$ consists of only one simple condition with the attribute whose typical (exceptional) value is sought, the "=" relational operator and a placeholder for the value sought. For example, in the context of personnel data, with $Q$ = *"most"* and $S$ = "age=?" (here "?" denotes a placeholder mentioned above), we look for a typical value of age. A Type 4 summary may produce typical (exceptional) values for some, possibly fuzzy, subset of rows. Type 5 summaries represent the most general form considered here: fuzzy rules describing dependencies between specific values of particular attributes. Here the use of $B$ is essential, while previously it was optional. The summaries of Type 1 and 3 have been implemented (Kacprzyk and Zadrożny, 2000b, 2000c, 2001c, 2005a) as an extension to Kacprzyk and Zadrożny's (1994, 1995a-b, 2001b) FQUERY for Access. Two approaches to Type 5 summaries generating has been proposed. Firstly, a subset of such summaries may be produced by using similarities with *association rules* (Agrawal and Srikant, 1994) and employing efficient algorithms for mining them. Second, genetic algorithm may be employed to search the space of possible summaries. We will refer the reader to the source Kacprzyk and Zadrożny's papers cited above.

Clearly, the protoforms are a powerful conceptual tool because we can formulate many different types of linguistic summaries in a uniform way, and devise a uniform and universal way to handle different linguistic summaries. Therefore, the use of protoforms is very relevant, and also contributes to an increased conceptual scalability of linguistic data summarization introduced by Kacprzyk and Zadrożny (2008a) as the simplicity and intuitive appeal of the protoforms used in the context of linguistic data summaries make them applicable to data sets of any size. Even if the size of a data set increases, the very essence of a particular protoform just catches the contents of the data set in a user comprehensible form.

Another aspect, which is relevant in our context, is whether one can also use in the process of linguistic summarization of data sets some other tools and techniques known in other areas, for which new, more effective and efficient approaches and methods are being proposed. If so, one could expect that we can use those new results for our ultimate benefit, that is, to make linguistic data summarization applicable to large problems.

In this perspective, it was shown in a recent paper by Kacprzyk and Zadrożny (2009b) that the linguistic data summarization as meant in this paper and viewed from the perspective of linguistic summaries as protoforms, is related to *natural language generation* (NLG).

Basically, natural language generation (NLG) is concerned with how one can automatically produce high quality natural language text from computer-internal representations of information which is not in natural language. In our case we follow the "numbers to words" path. NLG may be viewed from many perspectives (cf. Reiter and Dale, 2000) and for our purposes it may be expedient to consider independently the *tasks* of generation and the *process* of generation. As for the tasks, one can identify three types of generator *tasks*: text planning, sentence planning, and surface realization. In relation to our approach to linguistic summarization, we are mainly concerned with the text planning aspect since our approach is protoform based. However, the use of other two phases would produce noticeably more advanced linguistic summaries which could be in a position to capture more of fine shades of meaning. It is however not clear how to accommodate these tasks within the simple and efficient, yet somewhat strict Yager's concept of a linguistic summary.

Generator *processes* can be classified along two dimensions: sophistication and expressive power, starting with inflexible canned methods and ending with maximally flexible feature combination methods. The simplest approach, *canned text systems*, is used in many applications, notably simpler software systems which simply print a string of words without any change (error messages, warnings, letters, etc.). The approach can be used equally easily for single-sentence and for multisentence text generation. These systems are simple yet not "intelligent" enough.

More sophisticated are *template systems* which are used when a text (e.g. a message) must be produced several times with slight alterations as, e.g., in form letters (some open fields are to be filled in). The template approach is used mainly for multisentence generation, particularly when texts are rather regular in structure such as some business reports (e.g. stock market reports). In principle, our approach is similar in spirit to template based systems. One can say that Zadeh's protoforms can be viewed as playing a similar role to templates. However, one should bear in mind that there is an enormous difference between them. Namely, the protoforms are much more general and may represent such a wide array of various "templates" that maybe it would, more proper, to call them "families of templates" or "meta-templates".

An extremely interesting extension of our linguistic summarization might be to follow the multisentence path, for instance employing McKeown's (1985) idea of dynamically nesting instances of some paragraphs. However, again, it is not clear how one can extend the simple one sentence, protoform based structure of summaries adopted in our approach to this case.

*Phrase based systems* employ what can be seen as generalized templates. In such systems, a phrasal pattern is first selected to match the top level of the input, and then each part of the pattern is expanded into a more specific phrasal pattern that matches some subpart of the input, etc. with the phrasal pattern replaced by one or more words. Phrase based systems can be powerful and robust, but are very hard to build beyond a certain size, because of difficulties in a correct specification of the phrasal interrelationships.

It seems that our approach to linguistic summarization can be viewed, from some perspective, as a very simple phrase based system. It should be also noted that since protoforms may form hierarchies, we can imagine that both the phrase and its subphrases can be properly chosen protoforms. The calculi of fuzzy linguistically quantified statements can be extended to handle such a hierarchic structure of phrases (statements) though, at the semantic level, the same difficulties as in the NLG approach, notably an ability to grasp multisentence summaries with their interrelations, remain.

*Feature-based systems* (Cole et al., 1996) represent, in a sense, the extreme of the generalization of phrases. In feature based systems, each possible minimal alternative of expression is represented by a single feature, and the generation proceeds by the incremental collection of features appropriate for each part of the input until the sentence is fully determined. Feature based systems are the most sophisticated generators.

To summarize these short discussion of relations between our protoform base approach to linguistic data summarization, and modern approaches and solutions employed in the field of natural language generation (NLG), as first indicated by Kacprzyk and Zadrożny (2009b), we can clearly see that we can find much inspiration from recent developments in natural language generation, notably in the adjusting of protoforms to what is comprehensible and/or commonly used in a specific domain by using some sentence and text planning tools.

From the point of view of this paper, there is another aspect that is crucial. Namely, linguistic data summaries can provide an extremely human consistent tools for extracting knowledge from relevant, usually large, sets of data that is the foundation for a data driven decision support system as outlined in Section 1. The form of knowledge is extremely well comprehensible by the human user because it is in a simple natural language form. This can be decisive for an easy implementation of a data driven decision support system. In fact, this marvelous property of linguistic data summaries have been one of main reasons for the success of an implementation for a sales decision support for a small computer retailer (cf. Kacprzyk, 1999; Kacprzyk and Strykowski, 1999a, b; Zadrożny and Kacprzyk, 2007). We will now briefly show the very essence of how linguistic data summaries are used in the former implementation.

## 3 An Example: Linguistic Data Summaries to Support Sales Decision Making of a Computer Retailer

The proposed data summarization procedure was implemented to support sales decision making of a computer retailer in Southern Poland (cf. Kacprzyk, 1999; Kacprzyk and Strykowski, 1999a, b).

Though the database is large, its basic structure, which is relevant for our presentation, may be limited to the one shown in Table 2.

**Table 2.** Structure of the database

| Attribute name | Attribute type | Description |
|---|---|---|
| Date | Date | Date of sale |
| Time | Time | Time of sale transaction |
| Name | Text | Name of the product |
| Amount (number) | Numeric | Number of products sold in the transaction |
| Price | Numeric | Unit price |
| Commission | Numeric | Commission (in %) on sale |
| Value | Numeric | Amount (number) × price of the product |
| Discount | Numeric | Discount (in %) for transaction |
| Group | Text | Product group to which the product belongs |
| Transaction value | Numeric | Value of the whole transaction |
| Total sale to customer | Numeric | Total value of sales to the customer in fiscal year |
| Purchasing frequency | Numeric | Number of purchases by customer in fiscal year |
| Town | Text | Town where the customer lives |

**Table 3.** Linguistic summaries expressing relations between the group of products and commission

| Summary |
|---|
| About ½ of sales of network elements is with a high commission |
| About ½ of sales of computers is with a medium commission |
| Much sales of accessories is with a high commission |
| Much sales of components is with a low commission |
| About ½ of sales of software is with a low commission |
| About ½ of sales of computers is with a low commission |
| A few sales of components is without commission |
| A few sales of computers is with a high commission |
| Very few sales of printers is with a high commission |

**Table 4.** Linguistic summaries expressing relations between the groups of products and times of sale

| Summary |
|---|
| About 1/3 of sales of computers is by the end of year |
| About ½ of sales in autumn is of accessories |
| About 1/3 of sales of network elements is in the beginning of year |
| Very few sales of network elements is by the end of year |
| Very few sales of software is in the beginning of year |
| About ½ of sales in the beginning of year is of accessories |
| About 1/3 of sales in the summer is of accessories |
| About 1/3 of sales of peripherals is in the spring period |
| About 1/3 of sales of software is by the end of year |
| About 1/3 of sales of network elements is in the spring period |
| About 1/3 of sales in the summer period is of components |
| Very few sales of network elements is in the autumn period |
| A few sales of software is in the summer period |

In the beginning, after some initialization, we provide some parameters concerning mainly: definition of attributes and the subject, definition of how the results should be presented, definition of parameters of the method (i.e. a genetic algorithm or, seldom, full search). Then, we initialize the search and obtain results shown in the tables to follow.

These are the most valid summaries, and they give the user much insight into relations between the attributes chosen, moreover they are simple and human consistent.

Notice that these summaries concern data from the company's own database. However, companies operate in an environment (economic, climatic, social, etc.), and aspects of this environment may be relevant because they may greatly influence the operation, economic results, etc. of a particular company. A notable example may

**Table 5.** Linguistic summaries expressing relations between the attributes: size of customer, regularity of customer (purchasing frequency), date of sale, time of sale, commission, group of product and day of sale

| Summary |
|---|
| Much sales on Saturday is about noon with a low commission |
| Much sales on Saturday is about noon for bigger customers |
| Much sales on Saturday is about noon |
| Much sales on Saturday is about noon for regular customers |
| A few sales for regular customers is with a low commission |
| A few sales for small customers is with a low commission |
| A few sales for one-time customers is with a low commission |
| Much sales for small customers is for non-regular customers |

**Table 6.** Linguistic summaries expressing relations between groups of products, time of sale, temperature, precipitation, and type of customers

| Summary |
|---|
| Very few sales of software is in hot days to individual customers |
| About 1/2 of sales of accessories is in rainy days on weekends by the end of the year |
| About 1/3 of sales of computers is in rainy days to individual customers |

here be the case of climatic data that can be fetched from some sources, for instance from paid or free climatic data services. The inclusion of such data may be implemented as shown in Kacprzyk and Zadrozny (2005b).

It is quite obvious that though such data are widely available because meteorological services are popular around the world, the Internet is the best source of such data. This is particularly true in the case of a small company that has limited funds for data, and also limited human resources to fetch such data.

Using the data from meteorological commercial (inexpensive) and academic (free) services available through the Internet, we have been able to extend the system of linguistic database summarization described above.

For instance, if we are interested in relations between group of products, time of sale, temperature, precipitation, and type of customers, the best linguistic summaries (of both our "internal" data from the sales database, and "external" meteorological data from an Internet service) are as shown in Table 6.

Notice that the use of external data gives a new quality to possible linguistic summaries. It can be viewed as providing a greater adaptability to varying conditions because the use of free or inexpensive data sources from the Internet makes it possible to easily and quickly adapt the form and contents of summaries to varying needs and interests. And this all is practically at no additional price and effort. A more elaborate concept of a decision support system taking into account an information context of the decision making process has been proposed recently by Kacprzyk and Zadrożny (2008).

## 4 Concluding Remarks

In this paper we presented how the conceptually and numerically simple concept of a fuzzy linguistic database summary can be a very powerful tool for gaining much insight into which relations exist within data that may be relevant for a particular business activity and related decision making. The use of linguistic summaries can be described as providing tools for the verbalization of data analysis (mining) results which, in addition to the more commonly used visualization e.g. via a GUI, graphical user interface), can contribute to an increased human consistency and ease of use. The form of knowledge derived is in a simple, easily comprehensible linguistic form which can be effectively and efficiently employed for supporting decision makers, in the case considered along the data driven decision support system paradigm.

We have also mentioned two new relevant aspects the analysis of which was initiated by the authors. First, in Kacprzyk and Zadrożny (2009b) an extremely relevant aspect of scalability of linguistic summarization of data was considered and a new concept of a conceptual scalability was introduced. This is crucial for being able to proceed to large applications involving even huge data sets. Second, in Kacprzyk and Zadrożny (2009b) it was indicated that linguistic data summarization in the sense considered here is closely related to some types of solutions used in natural language generation (NLG), an area that is rapidly developing. Therefore, one can use more and more effective and efficient tools and techniques developed in that area to easier and faster derive even more comprehensible and up to the point linguistic data summaries.

We are convinced that linguistic data summaries will play a more and more relevant role in supporting human decision makers while solving difficult real life problems.

## References


Alter, S.L.: Decision Support Systems: Current Practice and Contributing Challenge. Addison-Wesley, Reading (1990)

Bosc, P., Dubois, D., Pivert, O., Prade, H., de Calmes, M.: Fuzzy summarization of data using fuzzy cardinalities. In: Proc. 7th Internat. Conf. IPMU (IPMU 2002), Annecy, France,, pp. 1553–1559 (2002)

Checkland, P.: Soft Systems Methodology. In: Rosenhead, J. (ed.) Rational Analysis for a Problematic World. John Wiley and Sons, Chichester (1989)

Checkland, P.B.: Soft systems methodology. A thirty year retrospective. In: Checkland, P.B., Scholes, J. (eds.) Soft Systems Methodology in Action. Wiley, Chichester (1999)

Cole, R.A., Mariani, J., Uszkoreit, H., Zaenen, A., Zue, V. (eds.): Survey of the State of the Art in Human Language Technology (1996), `http://cslu.cse.ogi.edu/HLTsurvey/HLTsurvey.html`

Dubois, D., Prade, H.: Fuzzy sets in data summaries–outline of a new approach. In: Proc. 8th Internat. Conf. IPMU (IPMU 2000), Madrid, Span, vol. 2, pp. 1035–1040 (2000)

George, R., Srikanth, R.: Data summarization using genetic algorithms and fuzzy logic. In: Herrera, F., Verdegay, J.L. (eds.) Genetic Algorithms and Soft Computing, pp. 599–611. Physica-Verlag, Heidelberg (1996)

Holsapple, C.W., Whinston, A.B.: Decision Support Systems: A Knowldege-based Approach. West Publishing Co., Minneapolis (1996)

Kacprzyk, J.: An Interactive Fuzzy Logic Approach to Linguistic Data Summaries. In: Proceedings of NAFIPS 1999 - 18th International Conference of the North American Fuzzy Information Processing Society – NAFIPS, pp. 595–599. IEEE, Los Alamitos (1999)

Kacprzyk, J., Strykowski, P.: Linguistic data summaries for intelligent decision support. In: Felix, R. (ed.) Fuzzy Decision Analysis and Recognition Technology for Management, Planning and Optimization – Proceedings of EFDAN 1999, pp. 3–12 (1999a)

Kacprzyk, J., Strykowski, P.: Linguistic Summaries of Sales Data at a Computer Retailer: A Case Study. In: Proceedings of IFSA 1999, Taipei, Taiwan R.O.C, vol. 1, pp. 29–33 (1999b)

Kacprzyk, J., Yager, R.R.: Linguistic summaries of data using fuzzy logic. Int. J. of General Systems 30, 133–154 (2001)

Kacprzyk, J., Yager, R.R., Zadrożny, S.: A fuzzy logic based approach to linguistic summaries of databases. Int. J. of Applied Mathematics and Computer Science 10, 813–834 (2000)

Kacprzyk, J., Zadrożny, S.: Fuzzy querying for Microsoft Access. In: Proceedings of FUZZ-IEEE 1994, Orlando, USA, vol. 1, pp. 167–171 (1994)

Kacprzyk, J., Zadrożny, S.: Fuzzy queries in Microsoft Access v. 2.0. In: Proceedings of FUZZ-IEEE/IFES 1995, Workshop on Fuzzy Database Systems and Information Retrieval, Yokohama, Japan, pp. 61–66 (1995a)

Kacprzyk, J., Zadrożny, S.: FQUERY for Access: fuzzy querying for a Windows–based DBMS. In: Bosc, P., Kacprzyk, J. (eds.) Fuzziness in Database Management Systems, pp. 415–433. Physica–Verlag, Heidelberg (1995b)

Kacprzyk, J., Zadrożny, S.: A fuzzy querying interface for a WWW–server–based relational DBMS. In: Proc. of IPMU 1996, Granada, Spain, vol. 1, pp. 19–24 (1996)

Kacprzyk, J., Zadrożny, S.: Data Mining via Linguistic Summaries of Data: An Interactive Approach. In: Yamakawa, T., Matsumoto, G. (eds.) Proceedings of IIZUKA 1998, Methodologies for the Conception, Design and Application of Soft Computing, Iizuka, Japan, pp. 668–671 (1998)

Kacprzyk, J., Zadrożny, S.: On Interactive Linguistic Summarization of Databases via a Fuzzy–Logic–Based Querying Add–On to Microsoft Access. In: Reusch, B. (ed.) Computational Intelligence: Theory and Applications, pp. 462–472. Springer, Heidelberg (1999)

Kacprzyk, J., Zadrożny, S.: On combining intelligent querying and data mining using fuzzy logic concepts. In: Bordogna, G., Pasi, G. (eds.) Recent Issues on Fuzzy Databases, pp. 67–81. Physica–Verlag, Heidelberg (2000a)

Kacprzyk, J., Zadrożny, S.: Data Mining via Fuzzy Querying over the Internet. In: Pons, O., Vila, M.A., Kacprzyk, J. (eds.) Knowledge Management in Fuzzy Databases, pp. 211–233. Physica–Verlag, Heidelberg (2000b)

Kacprzyk, J., Zadrożny, S.: Computing with words: towards a new generation of linguistic querying and summarization of databases. In: Sinčak, P., Vaščak, J. (eds.) Quo Vadis Computational Intelligence?, pp. 144–175. Physica-Verlag/Springer, Heidelberg (2000c)

Kacprzyk, J., Zadrożny, S.: Data mining via linguistic summaries of databases: an interactive approach. In: Ding, L. (ed.) A New Paradigm of Knowledge Engineering by Soft Computing, pp. 325–345. World Scientific, Singapore (2001a)

Kacprzyk, J., Zadrożny, S.: Computing with words in intelligent database querying: standalone and Internet-based applications. Information Sciences 34, 71–109 (2001b)

Kacprzyk, J., Zadrożny, S.: Fuzzy linguistic summaries via association rules. In: Kandel, A., Last, M., Bunke, H. (eds.) Data Mining and Computational Intelligence, pp. 115–139. Physica–Verlag/Springer, Heidelberg (2001c)

Kacprzyk, J., Zadrożny, S.: Protoforms of linguistic data summaries: towards more general natural–language–based data mining tools. In: Abraham, A., Ruiz del Solar, J., Koeppen, M. (eds.) Soft Computing Systems, pp. 417–425. IOS Press, Amsterdam (2002)

Kacprzyk, J., Zadrożny, S.: Linguistic database summaries and their protoforms: towards natural language based knowledge discovery tools. Information Sciences 173(4), 281–304 (2005a)

Kacprzyk, J., Zadrożny, S.: Towards a synergistic combination of Web–based and data–driven decision–support systems via linguistic data summaries. In: Szczepaniak, P.S., Kacprzyk, J., Niewiadomski, A. (eds.) AWIC 2005. LNCS (LNAI), vol. 3528, pp. 211–217. Springer, Heidelberg (2005b)

Kacprzyk, J., Zadrożny, S.: On a concept of a consensus reaching process support system based on the use of soft computing and Web techniques. In: Ruan, D., Montero, J., Lu, J., Martínez, L., D'hondt, P., Kerre, E.E. (eds.) Computational Intelligence in Decision and Control, pp. 859–864. World Scientific, Singapore (2008)

Kacprzyk, J., Zadrożny, S.: Linguistic data summarization: a high scalability through the use of natural language? In: Laurent, A., Lesot, M.-J. (eds.) Scalable Fuzzy Algorithms for Data Management and Analysis: Methods and Design (to appear 2009a)

Kacprzyk, J., Zadrożny, S.: Protoforms of Linguistic Database Summaries as a Human Consistent Tool for Using Natural Language in Data Mining. International Journal of Software Science and Computational Intelligence 1(1) (to appear, 2009b)

McKeown, K.R.: Text generation: using discourse strategies and focus constraints to generate natural language text. Cambridge University Press, Cambridge (1985)

Raschia, G., Mouaddib, N.: SAINTETIQ: a fuzzy set–based approach to database summarization. Fuzzy Sets and Systems 129, 137–162 (2002)

Rasmussen, D., Yager, R.R.: Using summarySQL as a tool for finding fuzzy and gradual functional dependencies. In: Proceedings of IPMU 1996, Granada, Spain, pp. 275–280 (1996)

Rasmussen, D., Yager, R.R.: A fuzzy SQL summary language for data discovery. In: Dubois, D., Prade, H., Yager, R.R. (eds.) Fuzzy Information Engineering: A Guided Tour of Applications, pp. 253–264. Wiley, New York (1997a)

Rasmussen, D., Yager, R.R.: SummarySQL–A fuzzy tool for data mining. Intelligent Data Analysis–An International Journal 1 (1997b) (Electronic Publication), `http://www-east.elsevier.com/ida/browse/96-6/ida96-6.htm`

Rasmussen, D., Yager, R.R.: Finding fuzzy and gradual functional dependencies with summarySQL. Fuzzy Sets and Systems 106, 131–142 (1999)

Reiter, E., Dale, R.: Building Natural Language Generation Systems. Cambridge University Press, Cambridge (2000)

Saaty, T.L.: The Analytic Hierarchy Process: Planning, Priority Setting, Resource Allocation. McGraw-Hill, New York (1980)

Sprague, R.H., Watson, H.J. (eds.): Decision Support Systems for Management. Prentice-Hall, Englewood Cliffs (1996)

Wierzbicki, A.P., Makowski, M., Wessles, J. (eds.): Model–based Decision Support Methodology with Environmental Applications. Kluwer, Dordrecht (2000)

Yager, R.R.: A new approach to the summarization of data. Information Sciences 28, 69–86 (1982)

Yager, R.R.: On ordered weighted averaging operators in multicriteria decision making. IEEE Trans. on Systems, Man and Cybern. SMC-18, 183–190 (1988)

Yager, R.R.: On linguistic summaries of data. In: Proceedings of IJCAI Workshop on Knowledge Discovery in Databases, Detroit, pp. 378–389 (1989)

Yager, R.R.: On linguistic summaries of data. In: Piatetsky-Shapiro, G., Frawley, B. (eds.) Knowledge Discovery in Databases, pp. 347–363. MIT Press, Cambridge (1991)

Yager, R.R.: Linguistic summaries as a tool for database discovery. In: Proceedings of Workshop on Flexible Query-Answering Systems, Roskilde University, Denmark, pp. 17–22 (1994)

Yager, R.R.: Linguistic summaries as a tool for database discovery. In: Proceedings of Workshop on Fuzzy Database Systems and Information Retrieval at FUZZ-IEEE/IFES, Yokohama, pp. 79–82 (1995a)

Yager, R.R.: Fuzzy summaries in database mining. In: Proceedings of the 11th Conference on Artificial Intelligence for Applications, Los Angeles, USA, pp. 265–269 (1995b)

Yager, R.R.: Database discovery using fuzzy sets. International Journal of Intelligent Systems 11, 691–712 (1996)

Yager, R.R., Kacprzyk, J. (eds.): The Ordered Weighted Averaging Operators: Theory and Applications. Kluwer, Boston (1997)

Yager, R.R., Rubinson, T.C.: Linguistic summaries of data bases. In: Proceedings of IEEE Conf. on Decision and Control, San Diego, USA, pp. 1094–1097 (1981)

Zadeh, L.A.: A computational approach to fuzzy quantifiers in natural languages. Computers and Maths with Appls. 9, 149–184 (1983)

Zadeh, L.A.: Syllogistic reasoning in fuzzy logic and its application to usuality and reasoning with dispositions. IEEE Transactions on Systems, Man and Cybernetics SMC–15, 754–763 (1985)

Zadeh, L.A.: A prototype-centered approach to adding deduction capabilities to search engines–the concept of a protoform. In: BISC Seminar, 2002. University of California, Berkeley (2002)

Zadeh, L.A.: From search engines to question answering systems–the problems of world knowledge relevance deduction and precisiation. In: Sanchez, E. (ed.) Fuzzy Logic and the Semantic Web, pp. 163–210. Elsevier, Amsterdam (2006)

Zadeh, L.A., Kacprzyk, J. (eds.): Computing with Words in Information/Intelligent Systems 1. Foundations. Physica–Verlag, Heidelberg (1999a)

Zadeh, L.A., Kacprzyk, J. (eds.): Computing with Words in Information/Intelligent Systems 2. Applications. Physica–Verlag, Heidelberg (1999b)

Zadrożny, S., Kacprzyk, J.: On database summarization using a fuzzy querying interface. In: Proceedings of IFSA 1999 World Congress, Taipei, Taiwan R.O.C., vol. 1, pp. 39–43 (1999)

Zadrożny, S., Kacprzyk, J.: Summarizing the contents of web server logs: a fuzzy linguistic approach. In: Proceedings of the 2007 IEEE Conference on Fuzzy Systems, pp. 100–105. Imperial College, London (2007)

# 9

# Computational Intelligence in Medical Decisions Making

Elizabeth Rakus-Andersson[1] and Lakhmi C. Jain[2]

[1]School of Engineering, Department of Mathematics and Science,
Blekinge University of Technology
Karlskrona, Sweden
Elisabeth.Andersson@bth.se

[2] School of Electrical and Information Engineering University of South Australia
Adelaide, Australia
Lakhmi.Jain@unisa.edu.au

**Abstract.** Computation intelligence paradigms including artificial neural networks, fuzzy systems, evolutionary computing techniques, intelligent agents and so on provide a basis for human like reasoning in medical systems. Approximate reasoning is one of the most effective fuzzy systems. The compositional rule of inference founded on the logical law modus ponens is furnished with a true conclusion, provided that the premises of the rule are true as well. Even though there exist different approaches to an implication, being the crucial part of the rule, we modify the early implication proposed in our practical model concerning a medical application. The approximate reasoning system presented in this work considers evaluation of a risk in the situation when physicians weigh necessity of the operation on a patient. The patient's clinical symptom levels, pathologically heightened, indicate the presence of a disease possible to recover by surgery. We wish to evaluate the extension of the operation danger by involving particularly designed fuzzy sets in the algorithm of approximate reasoning.

**Keywords:** Computational intelligence, approximate reasoning, compositional rule of inference, operation risk, symptom levels, parametric membership functions.

## 1 Introduction

Recent advances in computational intelligence techniques have offered tremendous opportunities to represent uncertain and imprecise knowledge in medical decision making. Artificial neural networks mimic the biological information processing mechanisms in a limited sense and help in fusing learning ability in decision making systems [1-11]. Fuzzy systems provide a means to transform computer programming to a sort of human reasoning system. Evolutionary computation involves a collection of algorithms based on the evolution of population towards a solution of a certain problem. Genetic algorithms, a part of evolutionary computing, are widely used in tasks such as optimization, automatic generation of artificial neural network architectures and so on. Multiagent systems are designed to act autonomously on behalf of the humans or users.

E. Rakus-Andersson et al. (Eds.): Recent Advances in Decision Making, SCI 222, pp. 145–159.
springerlink.com 

The tremendous interest in the applications of computational intelligence in healthcare among researchers is evident by a number of publications in journals and conference proceedings. A decision support system for breast cancer detection using Bayesian networks is reported in [12]. The authors have used multiple images of each breast and demonstrated the merit of their approach in comparison to the single image system. The use of the personalized reasoning mechanism for an intelligent medical e-learning system on atheromatosis is reported in [13]. Atheromatosis of the aortic arch has been recognized as an important source of embolism, which is a frequent cause of stroke. A new classification technique of continuous EEG recordings based on a network of spiking neurons is presented in [14]. The merit of the proposed technique is demonstrated.

Decision support systems in wireless capsule endroscopy are revisited in [15]. The author has reported a study on pattern recognition system for texture characterisation and classification of capsule-endoscopic images. A decision support scheme for choosing antibiotic in open heart surgery is presented in [16]. Methods such as kernel density estimation, with market basket analysis and text analysis for compression are used in the implementation of decision support system.

Fuzzy rule induction and artificial immune systems in breast cancer familiarity profiling is reported in [17]. It is demonstrated that the biologically inspired data mining techniques are competitive tools in cancer research. Online analytical process methodology for assessing the risk of developing acute coronary syndromes is reported in [18]. It is demonstrated that the technique offers a more accurate risk assessment as it takes into account variable interaction. A fuzzy relational based medical diagnostic decision support system is reported in [19]. The authors claim that the system replicates closely a physician's perception of symptom-disease associations and his/her approximate reasoning for diagnosis. An online decision support system for diagnosing hematologic malignancies by flow cytometry immunophenotyping is reported in [20]. The system is expected to facilitate clinical diagnosis of hematologic disease. A multiagent based healthcare system is reported in [21]. The system is aimed to help telemedicine service, patient monitoring and diagnosis, emergency management, and so on.

Mobile collaboration framework for u-healthcare agent services is presented in [22]. The authors have demonstrated successfully the use of agents in healthcare applications. The use of mobile agents for diagnostic support in ubiquitous healthcare is reported in [23]. The authors have proposed a mobile agent for diagnostic support by using neuro-fuzzy algorithm for consultation report. The merit of the proposed system is demonstrated. Privacy-aware autonomous agent system for pervasive healthcare is reported in [24]. The authors have developed the system which takes into account contexural information such as the user's location and identify, the time of the day, the artifacts used, and the presence of colleagues to infer hospital worker's availability and privacy demands. A new paradigm for modelling illness in the human population is reported in [25]. The authors have reported a patient model using a mobile software agent. It is believed that the patients can investigate the effects of their life styles on their medical conditions. A hybrid intelligent medical diagnostic system using the fusion of fuzzy and evolutionary algorithms is reported in [26]. The system is designed to diagnose and prescribe treatment of blood gas disturbances and disorders. The diagnosis process is modelled using domain expert and existing literature.

A diagnostic support system for bladder tumor grading is reported in [27]. The authors have combined fuzzy cognitive maps with support vector machines to achieve better tumor malignancy classification. The proposed system presents better classification accuracy than the existing systems and thus able to make decisions with high diagnostic accuracy. A computer aided diabetes management system is reported in [28]. A good review of the computer aided diabetes education e-learning tool and an approach to therapy management is presented by the author. The Glucose-Insulin and Glycemic Index Web Simulator (GIGISim) tool deals with the patient needs. Rule-based assistance to Brain Tumor Diagnosis is presented in [29]. The authors have used a brain tumor database consisting of nuclear magnetic resonance spectroscopic signals. It is demonstrated that three spectral frequencies are sufficient to represent to diagnose human brain tumor. The design of a learning environment for improving critical thinking skills in nursing domains is reported in [30]. The authors have presented the analysis of critical thinking. A learning system is proposed for facilitating decision making process. The system is in continuous improvement phase.

It is obvious from the above discussion that computational intelligence paradigms have become integral part of medical decision making. As a case study, we present the application of approximate reasoning in surgical decision making. The technique of approximate reasoning, earlier evolved by Zadeh [31, 32] quickly found many adherents who differentiated the foundations of the theory. Especially, the changes concerned the implication IF…THEN…ELSE…, which constitutes an important factor of the reasoning system. In [33-35] we can trace the discussion revealing definitions of the implication generated by Kleene and Denies, Willmot, Mamdani and Assilian, Larsen, Gödel et al. The trials of inserting individually created operations on fuzzy sets discern the approaches mentioned above. Even the item of compositional rule of inference was debated from separate points of views [36-40]. We can mention the Yager conception [40] and the Sugeno design [33] as the most original modifications of the initial version of the rule.

For a practitioner an applicable meaning of approximate reasoning is essential, especially in technique and natural sciences where vagueness of input and output is often expected. Although some technical trials of applications are remarkable, it can happen coincidentally to counterpart the approximate reasoning in medicine. The only contribution in the topic, found by the author in [31], is a discussion of the model employing a pharmacological example.

Since members of surgical staff make decisions about operations on severely-ill patients with the highest care then we wish to support these verdicts by results coming from reasoning systems. We adopt Zadeh's approach to the rule [31, 32, 42], which is slightly modified by us and based on Lukasiewicz's definition of the fuzzy implication [31, 36, 42]. We still find this rule to be the most appealing for the reason of simply performed operations and clearly interpretable results. Then we build an own original apparatus accommodated to medical assumptions. Particular fuzzy sets that contain input data and output effects are designed in compliance with the physician's hint. The discussion about how to find the objective of reasoning, i.e. operation risk, is accomplished in Section 2. Fuzzy sets, taking place in the model, are furnished with appropriate membership degrees in Section 3. Section 4, added as a presentation of efficiency of the algorithm, reveals some risks in cancer surgery.

## 2 Adoption of Approximate Reasoning to Operation Decisions

For patients, who suffer from e.g. cancer, decisions concerning their operations are made with the highest thoughtfulness. In the later or the last stage of the disease the possibility to cure the patient totally of cancer by operating him/her for tumors is rather little. As a physician does not want the patient to run the risk to suffer even more after an unnecessary operation, he ought to judge thoroughly the consequences of the surgery.

We intend to involve approximate reasoning to support mathematically the extraction of a proper decision when discerning the operation danger. The most decisive clinical symptoms found in an individual patient will be taken into consideration to evaluate the risk.

Let us ponder a logical compound statement

$$\begin{array}{l}\text{IF}\,(p\,\text{AND}\,((\text{IF}\,p\,\text{THEN}\ q)\,\text{ELSE}\,(\text{IF}\,(\text{NOT}\ p)\\ \text{THEN}\,(\text{NOT}\,q)))\,)\,\text{THEN}\ q\end{array} \tag{1}$$

whose primitive statements $p$ and $q$ are included in the equivalent form of (1) derived as

$$p \wedge ((p \rightarrow q) \wedge (\neg p \rightarrow \neg q)) \rightarrow q. \tag{2}$$

The logical joint ELSE is interpreted in (2) as the conjunction $\wedge$ in compliance with the suggestions made by Lukasiewicz and Zadeh [31, 36].

The logical statement (2) is a tautology, which can be easily confirmed by the method of truth tables. We also prove that thesis $q$ in (2) will become true if the premises $p$ and $(p \rightarrow q) \wedge (\neg p \rightarrow \neg q)$ constitute true statements as well. In order to accomplish the last proof we utilize the method of denying the truth of the thesis $q$. Let $v(p)$ and $v(q)$ denote the truth values of $p$ and $q$ according to the convention of binary logic. If, on behalf of the proof, we assume that the thesis $q$ is not true then $v(q) = 0$. From the previous assumption $v((p \rightarrow q) \wedge (\neg p \rightarrow \neg q)) = 1$ if $v(p \rightarrow q) = 1$ and $v(\neg p \rightarrow \neg q) = 1$. But $v(q) = 0$, which suggests that $v(p) = 0$ as well to warrant $v(p \rightarrow q) = 1$. On the other hand we have already assumed that premise $p$ is true. As the suggestion $v(q) = 0$ leads to the contradiction "$p$ is false" against "$p$ is true" then we will accept $v(q) = 1$.

In accordance with the extended law *modus ponens* proposed by Zadeh [31, 32] we interpret (2) as a sentence

IF
$p`$ (premise)
AND (3)

(IF $p$ THEN $q$) ELSE (IF (NOT $p$) THEN (NOT $q$)) (premise)
THEN
$q$` (thesis)

provided that the semantic meaning of $p$ and $p$` ($q$ and $q$` respectively) is very close.

Let $p$ be visualized by a fuzzy set $P$ in the universe $X$ and let $q$ be expressed by another fuzzy set $Q$ in the universe of discourse $Y$. Analogously, the fuzzy set $P`\subset X$ constitutes a mathematical formalization of the primitive statement $p$` whereas $Q`\subset Y$ replaces formally the sense of $q$`. The modus ponens rule thus becomes

IF
$p` = P`$ (premise)
AND (4)
(IF $p = P$ THEN $q = Q$) ELSE (IF (NOT $p = CP$) THEN (NOT $q = CQ$)) (premise)
THEN
$q`= Q`$ (thesis)

The sets $CP$ and $CQ$ are complements of $P$ and $Q$.

When making a feedback to the medical task previously outlined, we wish to use a technique of accommodating actual theoretical assertions to concrete formulations letting us evaluate the operation decision in some grades of risk.

Let $S$ denote a symptom possessing the most decisive power in the evaluation of the operation risk. We regard $S$ as either the complex qualitative symptom or the symptom whose intensity is assimilated with level codes. These codes, determined for both descriptions of $S$'s complexion, form the universe $X$ = "*symptom levels*" = $\{1,\ldots,k,\ldots,n\}$. Let us assume that level 1 is associated with the slightly heightened symptom values whereas level $n$ indicates their critical status.

The statement $p$`

$p$` = "symptom $S$ is found in patient on level $k$"

is now addressed to a fuzzy set $P$` introduced by

$$P` = \frac{\mu_{P`}(1)}{1} + \cdots + \frac{\mu_{P`}(k)}{k} + \cdots + \frac{\mu_{P`}(n)}{n}. \tag{5}$$

The sentence $p$ built by

"$p$ = "rising levels of $S$ are essential for operation risk"

is dedicated to a fuzzy set $P$ given by

$$P = \frac{\mu_P(1)}{1} + \cdots + \frac{\mu_P(k)}{k} + \cdots + \frac{\mu_P(n)}{n}. \tag{6}$$

Another category of elements, constituting a content of the universe $Y$, is determined in the model as risk grades. We set risk grades in $Y$ = "*operation risk grades*" = $\{L_0$ = "*none*", $L_1$ = "*little*", $L_2$ = "*moderate*", $L_3$ = "*great*", $L_4$ = "*total*"$\}$, on condition that $Y$ is experimentally restricted to five risk grades only.

For sentence $q$

$q$ = "operation risk exists for patient"

a creation of a fuzzy set $Q$ is supported by

$$Q = \frac{\mu_Q(L_0)}{L_0} + \frac{\mu_Q(L_1)}{L_1} + \frac{\mu_Q(L_2)}{L_2} + \frac{\mu_Q(L_3)}{L_3} + \frac{\mu_Q(L_4)}{L_4}. \tag{7}$$

At last, we define $q`$ containing the final risk judgment as a statement

$q`$ = "patient runs estimated risk of being operated",

where *risk* is graded by membership degrees of the corresponding fuzzy set $Q´$ proposed as

$$Q` = \frac{\mu_{Q`}(L_0)}{L_0} + \frac{\mu_{Q`}(L_1)}{L_1} + \frac{\mu_{Q`}(L_2)}{L_2} + \frac{\mu_{Q`}(L_3)}{L_3} + \frac{\mu_{Q`}(L_4)}{L_4}. \tag{8}$$

In the next paragraph we accomplish the discussion about an apparatus providing us with membership degrees of sets (5)–(8).

Due to *modus ponens* rule (4) we set all decision data in the scheme

IF
"symptom $S$ is found in patient on level $k$" = $P`$ (premise)
AND
(IF "rising levels of $S$ are essential for operation risk" = $P$ THEN "operation risk exists for patient" = $Q$) ELSE (IF ("rising levels of $S$ are not essential for operation risk" = $CP$ THEN operation risk does not exists for patient = $CQ$) (premise)
THEN
"patient runs estimated risk of being operated" = $Q`$ (thesis)

In conformity with [31, 36, 42] we first prognosticate a mathematical expression of the implication

(IF "rising levels of $S$ are essential for operation risk" = $P$ THEN "operation risk exists for patient" = $Q$) ELSE (IF ("rising levels of $S$ are not essential for operation risk" = $CP$ THEN operation risk does not exists for patient = $CQ$)

performed as matrix $R$. Even though several approaches to membership functions of implications were made [31-33, 35, 36, 38, 42] we still feel attracted by the Lukasiewicz [36, 42] conception of fuzzy implication $R$ with a membership function derived as

$$\begin{aligned} \mu_R(k, L_l) = 1 \wedge ((1 - \mu_P(k)) + \mu_Q(L_l)) \\ \wedge (\mu_P(k) + (1 - \mu_Q(L_l))), \end{aligned} \tag{9}$$

$k = 1, \ldots, n$, $l = 0, \ldots, 4$, for all $x \in X$ and all $y \in Y$.

The membership degrees of set $Q`$ will be visualized after composing set $P`$ with relation $R$ due to Zadeh's compositional rule [1]

$$Q` = P' \circ R \tag{10}$$

designated by the membership function

$$\mu_{Q`}(L_l) = \max_{k \in X}(\min(\mu_{P`}(k), \mu_R(k, L_l))) . \tag{11}$$

The comparisons of magnitudes of membership degrees in set $Q`$ yield indications referring to judgments of the risk grades after consideration of symptom level $k$ verified in the patient.

As the operations of maximum and minimum have a tendency to filter the input data, which sometimes does not result in a clear-cut decision, then we will propose another set of composition operations in (10). In accordance with [43] we propose

$$Q` = P' \underset{\cdot +}{\circ} R \tag{12}$$

assisted by membership degrees

$$\mu_{Q`}(L_l) = \frac{\sum_{k=1}^{n} \mu_{P`}(k) \cdot \mu_R(k, L_l)}{\sum_{k=1}^{n} \mu_R(L_l)} . \tag{13}$$

To be able to apply (13) we ought to prove that the value of the quotient $\mu_{Q`}(L_l)$ is a number belonging to the interval [0, 1]. To verify this we first notice that $\mu_{P`}(k) \cdot \mu_R(k, L_l) \le \mu_R(k, L_l)$ since both $\mu_{P`}(k)$ and $\mu_R(k, L_l)$ are less than one for all $k$ and $l$, $k = 1,\dots,n$, $l = 0,\dots,4$. This causes the value of a product to be lesser than the values of both factors. We thus conclude that the numerator is less than or equal to the denominator, which guarantees that the entire value of the quotient is a member of [0, 1]; therefore it can be approved as a membership degree of $L_l$ coming from the support of $Q`$.

We also notice that the sum placed in the denominator of the quotient never becomes equal to zero, since almost all risk grades will be designed as positive quantities. This assumption prohibits membership degrees of the risk grades from being undefined structures.

Values $\mu_R(k, L_l)$ are adaptable to be treated as weights of level importance assigned to a distinct risk. These, as the entries of matrix $R$ are invariants in the system promoting the same diagnostic model, contrary to information concerning different patients that is changeable. And, additionally, we can prove that operation (13) satisfies the criteria of OWA operators [43].

## 3 Mathematical Design of Data Sets

The decision model designed in Section 2 includes operations on fuzzy sets furnished with symbolically established membership degrees. In the current paragraph we put some life into theoretical symbols by assigning to them mathematical structures. The set $P`$ a.k.a. (5) now gets assigned

$$\begin{aligned} P` &= \frac{\mu_{P`}(1)}{1} + \dots + \frac{\mu_{P`}(k)}{k} + \dots + \frac{\mu_{P`}(n)}{n} \\ &= \dots + \frac{\frac{n-2}{n}}{k-2} + \frac{\frac{n-1}{n}}{k-1} + \frac{1}{k} + \frac{\frac{n-1}{n}}{k+1} + \frac{\frac{n-2}{n}}{k+2} + \dots \end{aligned} \tag{14}$$

for the $k^{\text{th}}$ symptom level certified in the patient examined.

Another set $P$, concerning the same symptom levels in the support, is found by (6) and modified as

$$\begin{aligned} P &= \frac{\mu_P(1)}{1} + \dots + \frac{\mu_P(k)}{k} + \dots + \frac{\mu_P(n)}{n} \\ &= \frac{\frac{1}{n}}{1} + \dots + \frac{\frac{k}{n}}{k} + \dots + \frac{\frac{n}{n}}{n}, \end{aligned} \tag{15}$$

due to the previously made assumptions, which suggest the tendency to ascending values of the membership degrees in $P$.

The set $Q$ is more sophisticated to design as a fuzzy set whose support consists of other fuzzy sets $L_l$, $l = 0,\dots,4$, commonly defined in a symbolic risk reference set $Z = [0, 1]$. We also intend to determine the membership degrees of $Q$ as some characteristic quantities from [0, 1]. Evaluation of these numbers is founded on a procedure involving a linguistic variable

"*operation risk grades*" = {$L_0$ = "*none*", $L_1$ = "*little*", $L_2$ = "*moderate*", $L_3$ = "*great*", $L_4$ = "*total*"},

experimentally restricted to five risk grades only.

We first fuzzify the expressions concerning the items of the list to continue further with their defuzzification in order to attach numerical equivalents to the words from the list. Each word assists now a fuzzy set $L_l$, $l$ = 0, 1, 2, 3, 4, whose constraint is grounded on an $s$-class mapping defined for $z$ in $Z$ = [0, 1] as [44]

$$\mu_{L_l}(z) = \mu_{L_0(l)}(z) = \begin{cases} \text{left}(\mu_{L_0(l)}(z)) = s(z, \alpha_{L_0}, \beta_{L_0}, \gamma_{L_0}, l \cdot h), & \text{for } z \le \gamma_{L_0}, \\ \text{right}(\mu_{L_0(l)}(z)) = 1 - s(z, \alpha_{L_0} + h, \beta_{L_0} + h, \gamma_{L_0} + h, l \cdot h) & \text{for } z > \gamma_{L_0}. \end{cases} \tag{16}$$

We clarify the fact that formulas of all membership functions are derived from only one predetermined subject defining $\mu_{L_0}(z)$. The equality $\mu_{L_l}(z) = \mu_{L_0(l)}(z)$ reveals that $\mu_{L_l}(z)$ is dependent on a parameter $l$ equal to level number $l$, $l = 0,\ldots,4$. The $h$ unit determines a distance between $\alpha_{L_l}$ and $\alpha_{L_{l+1}}$ (respectively $\beta_{L_l}$ and $\beta_{L_{l+1}}$ or $\gamma_{L_l}$ and $\gamma_{L_{l+1}}$) for symmetric functions $s$.

We prepare constraints for $L_0$, which are affected by $\alpha_{L_0} = -0.25$, $\beta_{L_0} = -0.125$ and $\gamma_{L_0} = 0$ as

$$\mathrm{left}(\mu_{L_0}(z)) = \begin{cases} 2\left(\frac{z-(-0.25)}{0-(-0.25)}\right)^2 \\ \text{for} \quad -0.25 \le z < -0.125, \\ 1-2\left(\frac{z-0}{0-(-0.25)}\right)^2 \\ \text{for} \quad -0.125 \le z < 0, \end{cases} \tag{17}$$

and

$$\mathrm{right}(\mu_{L_0}(z)) = \begin{cases} 1-2\left(\frac{z-0}{0.25-0}\right)^2 \\ \text{for} \quad 0 \le z < 0.125, \\ 2\left(\frac{z-0.25}{0.25-0}\right)^2 \\ \text{for} \quad 0.125 \le z < 0.25. \end{cases} \tag{18}$$

By inserting in (17) and (18) the current value $l$, $l = 0,\ldots,4$, and the distance $h$, casually determined as $h = 0.25$, we obtain a formula of the left branch of $L_l$

$$\mathrm{left}(\mu_{L_l}(z)) = \begin{cases} 2\left(\frac{z-(-0.25+l\cdot 0.25)}{0-(-0.25)}\right)^2 \\ \text{for} \ -0.25+l\cdot 0.25 \le z < -0.125+l\cdot 0.25, \\ 1-2\left(\frac{z-(0+l\cdot 0.25)}{0-(-0.25)}\right)^2 \\ \text{for} \ -0.125+l\cdot 0.25 \le z < 0+l\cdot 0.25, \end{cases} \tag{19}$$

and a function shaping its right branch

$$\mathrm{right}(\mu_{L_l}(z)) = \begin{cases} 1-2\left(\frac{z-(0+l\cdot 0.25)}{0.25-0}\right)^2 \\ \text{for} \ 0+l\cdot 0.25 \le z < 0.125+l\cdot 0.25, \\ 2\left(\frac{z-(0.25+l\cdot 0.25)}{0.25-0}\right)^2 \\ \text{for} \ 0.125+l\cdot 0.25 \le z < 0.25+l\cdot 0.25. \end{cases} \tag{20}$$

Figure 1 collects plots of $L_0$–$L_4$ in conformity with different values of $l$ included in (19) and (20).

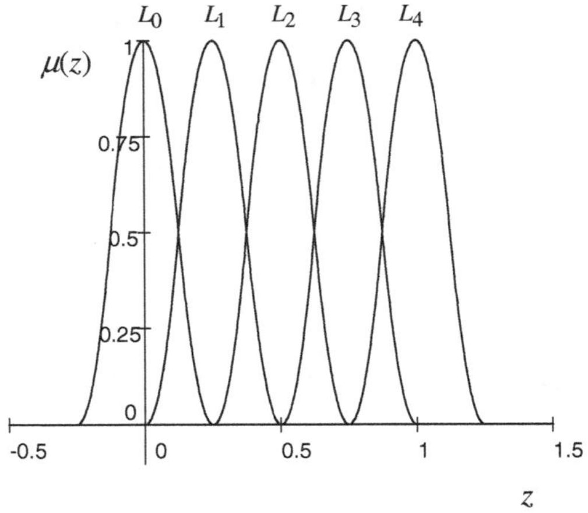


**Fig. 1.** The terms of "*operation risk grades*" as fuzzy sets $L_0$–$L_4$

Actually, we have an intension to emphasize the meaning of parametric nature of the $L_l$ membership functions, which deprives the model of many distinct formulas. Apart from this advantage we focus on generating the functions that represent elegant structures mathematically expressed.

In the process of defuzzification we consider only $z$-values for which the sets $L_0$–$L_4$ get the status of normal sets, i.e., $z = 0$, $z = 0.25$, $z = 0.5$, $z = 0.75$ and $z = 1$. For these, another fuzzy set "*numerical operation risk*" is projected by developing its membership function in the form of

$$\mu_{\text{"numerical operation risk"}}(z) = \begin{cases} 2\left(\frac{z-0}{1-0}\right)^2 \\ \text{for } 0 \le z < 0.5, \\ 1 - 2\left(\frac{z-1}{1-0}\right)^2 \\ \text{for } 0.5 \le z < 1. \end{cases} \tag{21}$$

Via the selected $z$-quantities above, we tie their membership degrees calculated by means of (21) to expressions from the list in order to establish relations between words and their numerical replacements. Therefore, the set $Q$ finally obtains a shape of

$$Q = \frac{0}{L_0} + \frac{0.125}{L_1} + \frac{0.5}{L_2} + \frac{0.875}{L_3} + \frac{1}{L_4}. \tag{22}$$

We now wish to demonstrate the action of approximate reasoning accustomed to the judgment of surgical risk.

## 4 Risks Grades in Cancer Surgery

In patients, who suffer from cancer as the recognized diagnosis, one of the symptoms, namely, *CRP* (*C*-reactive proteins) is carefully measured and discussed with a view to make a decision about accomplishing a successful operation. The heightened values of *CRP* (measured in milligrams per liter) are theoretically discerned in four levels stated as

1 = "*almost normal*" for $CRP < 10$,
2 = "*heightened*" if $10 \le CRP \le 20$,
3 = "*very heightened*" if $20 \le CRP \le 25$,
4 = "*dangerously heightened*" for $CRP > 25$.

Due to (15) set $P$ is expressed as

$$P = \frac{0.25}{1} + \frac{0.5}{2} + \frac{0.75}{3} + \frac{1}{4} \tag{23}$$

in $X = \{1,\ldots,4\}$.

Suppose that an individual patient examined reveals the *CRP*-value to be 23. *CRP* is thus classified in level 3 and set $P`$ characteristic of the patient is stated in the form of

$$P` = \frac{0.5}{1} + \frac{0.75}{2} + \frac{1}{3} + \frac{0.75}{4} \ . \tag{24}$$

according to (14).

The sets (23) and (22) together with

$$CP = \frac{0.75}{1} + \frac{0.5}{2} + \frac{0.25}{3} + \frac{0}{4} \tag{25}$$

and

$$CQ = \frac{1}{L_0} + \frac{0.875}{L_1} + \frac{0.5}{L_2} + \frac{0.125}{L_3} + \frac{0}{L_4} \tag{26}$$

generate matrix $R$ with the entries computed in compliance with (9). $R$ is expanded as a two-dimensional table

$$R = \begin{array}{c} \\ 1 \\ 2 \\ 3 \\ 4 \end{array} \begin{array}{c} \begin{array}{ccccc} L_0 & L_1 & L_2 & L_3 & L_4 \end{array} \\ \begin{bmatrix} 0.75 & 0.875 & 0.75 & 0.375 & 0.25 \\ 0.5 & 0.625 & 1 & 0.625 & 0.5 \\ 0.25 & 0.375 & 0.75 & 0.875 & 0.75 \\ 0 & 0.125 & 0.5 & 0.875 & 1 \end{bmatrix} \end{array} \tag{27}$$

which, inserted in (10) for $P`$ determined by (24), provides us with

$$Q` = \frac{0.5}{L_0} + \frac{0.625}{L_1} + \frac{0.75}{L_2} + \frac{0.875}{L_3} + \frac{0.75}{L_4} \ . \tag{28}$$

By interpreting the meaning of (28) we understand that there exists a risk when considering an operation in patient whose *CRP*-index is evaluated on the third level. The most possible risk is evaluated as "*great*" according to the highest quantity of the membership degree. The total danger of accomplishing the surgical operation is evaluated as essential with the membership degree 0.75.

Even the results of implementing (13) given as

$$Q`=\frac{0.66}{L_0}+\frac{0.69}{L_1}+\frac{0.75}{L_2}+\frac{0.795}{L_3}+\frac{0.725}{L_4}. \quad (29)$$

fully confirm the risk extension judged by (28).

We hope that the classical model of approximate reasoning, modified by us and adapted to the problem of operation decision can constitute its complementary solution, especially when a decision of saving somebody's life via surgery is crucial.

## 5 Conclusions

We have presented an overview of the computational intelligence paradigms in medical decision making. As a case study, we have used approximated reasoning to introduce the initial interpretation of the system to approximate the operation risk concerning patients with rising values of a biological index. The formulas of membership degrees and membership functions have been expanded by applying a formal mathematical design. We expect that the study makes a contribution in the domain of mathematical models projected for medical applications.

In future works we wish to examine a model consisted of several symptoms that are divided in different numbers of levels. The symptoms should be included in the pattern simultaneously, which may expose some internal interactions among them. In other words, the operation risk will be a criterion that can employ many data factors. We count on finding some helpful remarks in [45] to implement an algorithm supporting the method newly planned.

## Acknowledgement

This chapter is based on Professor Rakus Andersson's research paper accepted for presentation in 2009 IFSA World Congress, Lisbon, Portugal.

## References

[1] Teodorescu, H.-N., Kandel, A., Jain, L.C. (eds.): Fuzzy and Neuro-fuzzy Systems in Medicine. CRC Press, USA (1999)

[2] Teodorescu, H.-N., Kandel, A., Jain, L.C. (eds.): Soft Computing in Human Related Sciences. CRC Press, USA (1999)

[3] Tonfoni, G., Jain, L.C. (eds.): Innovations in Decision support Systems, Advanced Knowledge International, ch. 8, 9 (2003)

[4] Jain, A., Jain, A., Jain, S., Jain, L.C. (eds.): Artificial Intelligence Techniques in Breast Cancer Diagnosis and Prognosis. World Scientific, Singapore (2000)
[5] Teodorescu, H.-N., Jain, L.C. (eds.): Intelligent Systems and Technologies in Rehabilitation Engineering. CRC Press, USA (2001)
[6] Silverman, B., Jain, A., Ichalkaranje, A., Jain, L.C. (eds.): Intelligent Paradigms in Healthcare Enterprises. Springer, Heidelberg (2005)
[7] Husmeier, D., Dybowski, R., Roberts, S. (eds.): Probabilistic Modelling in Bioinformatics and Medical Informatics. In: Wu, X., Jain, L.C. (eds.) Advanced Information and Knowledge Processing Series. Springer, Heidelberg (2005)
[8] Ichalkaranje, N., Ichalkaranje, A., Jain, L.C. (eds.): Intelligent Paradigms for Assistive and Preventive Healthcare. Springer, Heidelberg (2006)
[9] Yoshida, H., Jain, A., Ichalkaranje, A., Jain, L.C., Ichalkaranje, N. (eds.): Advanced Computational Intelligence Paradigms in Healthcare 1. Springer, Heidelberg (2007)
[10] Vaidya, S., Jain, L.C., Yoshida, H. (eds.): Advanced Computational Intelligence Paradigms in Healthcare 2. Springer, Heidelberg (2008)
[11] Sardo, M., Vaidya, S., Jain, L.C. (eds.): Advanced Computational Intelligence Paradigms in Healthcare 3. Springer, Heidelberg (2008)
[12] Velikova, M.: A Decision support system for Breast Cancer Detection in Screening Programs. In: Proceedings of the 18th European Conference on Artificial Intelligence, pp. 658–662. IOS Press, Amsterdam (2008)
[13] Kabassi, K., et al.: Specifying the personalization reasoning mechanism for an intelligent medical e-learning system on Atheromatosis: An Empirical Study. Intelligent Decision Technologies-An International Journal 2(3), 179–190 (2008)
[14] Goel, P., et al.: On the Use of Spiking Neural Networks for EEG Classification. International Journal of Knowledge-Based and Intelligent Engineering Systems 12(4), 295–304 (2008)
[15] Kodogiannis, V.S.: Decision Support Systems in Wireless Capsule Endoscopy. Intelligent Decision Technologies-An International Journal 1(1-2), 17–32 (2007)
[16] Cerrito, P.B.: Choice of Antibiotic in Open Heart surgery. Intelligent Decision Technologies-An International Journal 1(1-2), 63–69 (2007)
[17] Menolascina, F., et al.: Fuzzy rule Induction and Artificial Immune Systems in Female Breast cancer Familiarity Profiling. International Journal of Hybrid Intelligent Systems 5(3), 161–165 (2008)
[18] Kostakis, H., et al.: A Computational Algorithm for the Risk assessment of Developing Acute Coronary Syndromes using Online Analytical Process Methodology. International Journal of Knowledge Engineering and Soft Data Paradigms 1(1), 85–99 (2009)
[19] Wagholikar, K.V., Deshpande, A.W.: Fuzzy Relation Based Modeling for Medical Diagnostic Decision Support: Case Studies. International Journal of Knowledge-Based and Intelligent Engineering Systems 12(5-6), 319–326 (2008)
[20] Qian, Y.–W., et al.: An On-line Decision Support System for Diagnosing Hematologic Malignancies by Flow Cytometry Immunophenotyping. International Journal of Medical Engineering and Informatics 1(1), 109–124
[21] Kang, E., et al.: Remote control multi-agent system for u-healthcare service. In: Nguyen, N.T., Grzech, A., Howlett, R.J., Jain, L.C. (eds.) KES-AMSTA 2007. LNCS (LNAI), vol. 4496, pp. 636–644. Springer, Heidelberg (2007)
[22] Jeong, C.-W., Kim, D.-H., Joo, S.-C.: Mobile collaboration framework for u-healthcare agent services and its application using pDAs. In: Nguyen, N.T., Grzech, A., Howlett, R.J., Jain, L.C. (eds.) KES-AMSTA 2007. LNCS (LNAI), vol. 4496, pp. 747–756. Springer, Heidelberg (2007)

[23] Mateo, R.M.A., et al.: Mobile Agent Using Data Mining for Diagnostic Support in Ubiquitous Healthcare. In: Nguyen, N.T., Grzech, A., Howlett, R.J., Jain, L.C. (eds.) KES-AMSTA 2007. LNCS (LNAI), vol. 4496, pp. 795–804. Springer, Heidelberg (2007)
[24] Tentori, M., et al.: Privacy-Aware Autonomous agents for Pervasive Healthcare. IEEE Intelligent Systems 21(6), 55–62 (2006)
[25] Nejad, S.G., et al.: An Agent-based Diabetic Patient Simulation. In: Nguyen, N.T., Jo, G.S., Howlett, R.J., Jain, L.C. (eds.) KES-AMSTA 2008. LNCS (LNAI), vol. 4953, pp. 832–841. Springer, Heidelberg (2008)
[26] Koutsojannis, C., Hatzillygeroudis, I.: Fuzzy-Evolutionary Synergism in an Intelligent Medical Diagnosis System. In: Gabrys, B., Howlett, R.J., Jain, L.C. (eds.) KES 2006. LNCS (LNAI), vol. 4252, pp. 1313–1322. Springer, Heidelberg (2006)
[27] Papageorgiou, E., et al.: Combining Fuzzy Cognitive Maps with Support Vector Machines for Bladder Tumor Grading. In: Gabrys, B., Howlett, R.J., Jain, L.C. (eds.) KES 2006. LNCS (LNAI), vol. 4251, pp. 515–523. Springer, Heidelberg (2006)
[28] Koleszynska, J.: GIGISim-The Intelligent Telehealth System: Computer Aided Diabetes Management – A New Review. In: Apolloni, B., Howlett, R.J., Jain, L. (eds.) KES 2007, Part I. LNCS (LNAI), vol. 4692, pp. 789–796. Springer, Heidelberg (2007)
[29] Nebot, A.: Rule-Based Assistance to Brain Tumor Diagnosis Using LR-FIR. In: Lovrek, I., Howlett, R.J., Jain, L.C. (eds.) KES 2008, Part II. LNCS (LNAI), vol. 5178, pp. 173–180. Springer, Heidelberg (2008)
[30] Seta, K., et al.: Learning Environment for Improving Critical Thinking Skills in Nursing Domain. International Journal of Advanced Intelligence Paradigms 1(2) (2008)
[31] Zadeh, L.A.: Outline of a new approach to the analysis of complex systems and decision process. IEEE Transactions on Systems, Man, and Cybernetics MSC-3(1), 28–44 (1973)
[32] Zadeh, L.A.: A theory of approximate reasoning. Machine Intelligence 9, 149–174 (1979)
[33] Fuller, R.: On fuzzy reasoning schemes, `http://www.abo.fi/~rfuller/pgs97.pdf`
[34] Hüllermeier, E.: Case based approximate reasoning. Springer, Heidelberg (2007)
[35] van den Broek, P.M.: Efficient algorithms for approximate reasoning. In: Proceedings of the 6th International Conference on Neural Information Processing, ICONIP 1999, Perth, Western Australia (1999)
[36] Baldwin, J.F., Pilsworth, B.W.: A model of fuzzy reasoning through multi-valued logic and set theory. Int. J. of Man-Machine Studies 11, 351–380 (1979)
[37] Cornelis, C., de Cock, M., Kerre, E.: Efficient approximate reasoning with positive and negative information. In: Negoita, M.G., Howlett, R.J., Jain, L.C. (eds.) KES 2004. LNCS (LNAI), vol. 3214, pp. 779–785. Springer, Heidelberg (2004)
[38] Gerla, G.: Fuzzy logic: mathematical tools for fuzzy reasoning. Kluwer, Dordrecht (2001)
[39] Mamdani, E.H., Assilian, S.: An experiment in linguistic synthesis with a fuzzy logic controller. Int. J. of Man-Machine Studies 7, 1–13 (1975)
[40] Yager, R.: An approach to inference in approximate reasoning. Int. J. of Man-Machine Studies 13, 323–338 (1980)
[41] Kilic, K., Sproule, B.A., Türksen, I.B., Naranjo, C.A.: A fuzzy system modeling algorithm for data analysis and approximate reasoning, `https://research.sabanciuniv.edu/418/1/robotics_kemal.DOC.R`
[42] Kacprzyk, J.: Fuzzy sets in system analysis. PWN, Warsaw (1986)

[43] Rakus-Andersson, E.: Fuzzy and rough techniques in medical diagnosis and medication. Springer, Heidelberg (2007)

[44] Rakus-Andersson, E.: S-truncated functions and rough sets in approximation and classification of polygons. In: Proc. of Modeling Decisions for Artificial Intelligence – MDAI 2005, Tsukuba, CD-ROM, paper nr 049, Consejo Superior de Investigaciones Cientificas (2005)

[45] Wolkenhauer, O.: Fuzzy inference engines, composition and individual rule based on non-linear mappings, `http://www.Systemsbiology.umist.ac.uk/dokumente/dataengineering/infengines.pdf`

# Appendix

Following is a sample of additional resources on intelligent systems and technologies.

## Journals

- International Journal of Knowledge-Based intelligent Engineering systems, IOS Press, The Netherlands.
  http://www.kesinternational.org/journal/
- International Journal of Hybrid Intelligent Systems, IOS Press, The Netherlands.
  http://www.iospress.nl/html/14485869.html
- Intelligent Decision Technologies: An International Journal, IOS Press, The Netherlands.
  http://www.iospress.nl/html/18724981.html
- IEEE Intelligent Systems, IEEE Press, USA.
  www.computer.org/intelligent/
- IEEE Transactions on Neural Networks.
- IEEE Transactions on Evolutionary Computing.
- IEEE Transactions on Fuzzy Systems.
- IEEE Computational Intelligence Magazine.
- Neural Computing and applications, Springer.
- Neurocomputing, Elsevier.
- International Journal of Intelligent and Fuzzy Systems, IOS Press, The Netherlands.
- Fuzzy Optimization and Decision Making, Kluwer.
- AI Magazine, USA
  www.aaai.org/

## Special Issue of Journals

- Nguyen, N.T., Lim, C.P., Jain, L.C. and Balas, V.E. (Guest Editors), Theoretical Advances and Applications of Intelligent Paradigms, Journal of Intelligent and Fuzzy Systems, IOS Press, Volume 20, Numbers 1,2, 2009.
- Jain, L.C., Lim, C.P. and Nguyen, N.T. (Guest Editors), Recent Advances in Intelligent Paradigms Fusion and Their Applications, International Journal of Hybrid Intelligent Systems, Volume 5, Issue 3, 2008.
- Lim, C.P., Jain, L.C., Nguyen, N.T. and Balas, V. (Guest Editors), Advances in Computational Intelligence Paradigms and Applications, An International Journal on Fuzzy Optimization and Decision Making, Kluwer Academic Publisher, in press.
- Abraham, A., Jarvis, D., Jarvis, J. and Jain, L.C. (Guest Editors), Special issue on Innovations in agents: An International Journal on Multiagent and Grid Systems, IOS Press, Volume 4, Issue 4, 2008.

- Ghosh, A., Seiffert, U. and Jain, L.C. (Guest Editors), Evolutionary Computation in Bioinformatics, Journal of Intelligent and Fuzzy Systems, IOS Press, The Netherlands, Volume 18, Number 6, 2007.
- Abraham, A., Smith, K., Jain, R. and Jain, L.C. (Guest Editors), Network and Information Security: A Computational Intelligence Approach, Journal of Network and Computer Applications, Elsevier Publishers, Volume 30, Issue 1, 2007.
- Palade, V. and Jain, L.C. (Guest Editors), Practical Applications of Neural Networks, Journal of Neural Computing and Applications, Springer, Germany, Volume 14, No. 2, 2005.
- Abraham, A. and Jain, L.C. (Guest Editors), Computational Intelligence on the Internet, Journal of Network and Computer Applications, Elsevier Publishers, Volume 28, Number 2, 2005.
- Abraham, A., Thomas, J., Sanyal, S. and Jain, L.C. (Guest Editors), Information Assurance and Security, Journal of Universal Computer Science, Volume 11, Issue 1, 2005.
- Abraham, A. and Jain, L.C. (Guest Editors), Optimal Knowledge Mining, Journal of Fuzzy Optimization and Decision Making, Kluwer Academic Publishers, Volume 3, Number 2, 2004.
- Palade, V. and Jain, L.C. (Guest Editors), Engineering Applications of Computational Intelligence, Journal of Intelligent and Fuzzy systems, IOS Press, Volume 15, Number 3, 2004.
- Alahakoon, D., Abraham, A. and Jain, L.C. (Guest Editors), Neural Networks for Enhanced Intelligence, Neural Computing and Applications, Springer, UK, Volume 13, No. 2, June 2004.
- Abraham, A., Jonkar, I., Barakova, E., Jain, R. and Jain, L.C. (Guest Editors), Special issue on Hybrid Neurocomputing, Neurocomputing, Elsevier, The Netherlands, Volume 13, No. 2, June 2004.
- Abraham, A. and Jain, L.C. (Guest Editors), Knowledge Engineering in an Intelligent Environment, Journal of Intelligent and Fuzzy Systems, IOS Press, The Netherlands, Volume 14, Number 3, 2003.
- Jain, L.C. (Guest Editor), Fusion of Neural Nets, Fuzzy Systems and Genetic Algorithms in Industrial Applications, IEEE Transactions on Industrial Electronics, USA, Volume 46, Number 6, December 1999.
- De Silva, C. and Jain, L.C. (Guest Editors), Intelligent Electronic Systems, Engineering Applications of Artificial Intelligence, Pergamon Press, USA, Volume 11, Number 1, January 1998.
- Jain, L.C. (Guest Editor), Intelligent Systems: Design and Applications - 2, Journal of Network and Computer Applications, Elsevier, **Vol. 2**, April 1996.
- Jain, L.C. (Guest Editor), Intelligent Systems: Design and Applications - 1, Journal of Network and Computer Applications, Elsevier, **Vol.1**, January, 1996.

## Conferences

- KES International Conference Series
  www.kesinternational.org/
- AAAI Conference on Artificial Intelligence
  www.aaai.org/aaai08.php
- European Conferences on Artificial Intelligence (ECAI)

## Conference Proceedings

- *Håkansson*, A., Nguyen, N.T., Hartung, R., Howlett, R.J. and Jain, L.C. (Editors), Agents and Multi-Agents Systems: Technologies and Applications, Lecture Notes in Artificial Intelligence, Springer-Verlag, Germany, 2009.
- Nakamatsu, K., Phillips-Wren, G., Jain, L.C. and Howlett, R.J. (Editors), New Advances in Intelligent Decision Technologies, Springer-Verlag, 2009.
- Nguyen, N.T., Jo, G.S., Howlett, R.J. and Jain, L.C. (Editors), Agents and Multi-Agents Systems: Technologies and Applications, Lecture Notes in Artificial Intelligence LNAI 4953, Springer-Verlag, Germany, 2008.
- Lovrek, I., Howlett, R.J. and Jain, L.C. (Editors), Knowledge-Based and Intelligent Information and Engineering Systems, Lecture Notes in Artificial Intelligence, Volume 1, LNAI 5177, Springer-Verlag, Germany, 2008.
- Lovrek, I., Howlett, R.J. and Jain, L.C. (Editors), Knowledge-Based and Intelligent Information and Engineering Systems, Lecture Notes in Artificial Intelligence, Volume 2, LNAI 5178, Springer-Verlag, Germany, 2008.
- Lovrek, I., Howlett, R.J. and Jain, L.C. (Editors), Knowledge-Based and Intelligent Information and Engineering Systems, Lecture Notes in Artificial Intelligence, Volume 3, LNAI 5179, Springer-Verlag, Germany, 2008.
- Pan, J. S., Niu, X.M., Huang, H. C. and Jain, L.C., Intelligent Information Hiding and Multimedia Signal Processing, IEEE Computer Society Press, USA, 2008.
- Jain, L.C., Lingras, P., Klusch, M., Lu, J., Zhang, C., Cercone, N. and Cao, L. Proceedings of the 2008 IEEE/WIC/ACM International Conference on Web Intelligence, WI 2008, IEEE Computer Society, USA, 2008.
- Jain, L.C., Gini, M., Faltings, B.B., Terano, T., Zhang, C., Cercone, N. and Cao, L. Proceedings of the 2008 IEEE/WIC/ACM International Conference on Intelligent Agent Technology, IAT 2008, IEEE Computer Society, USA, 2008.
- Apolloni, B., Howlett, R.J. and Jain, L.C. (Editors), Knowledge-Based Intelligent Information and Engineering Systems, Lecture Notes in Artificial Intelligence, Volume 1, LNAI 4692, KES 2007, Springer-Verlag, Germany, 2007.
- Apolloni, B., Howlett, R.J.and Jain, L.C. (Editors), Knowledge-Based Intelligent Information and Engineering Systems, Lecture Notes in Artificial Intelligence, Volume 2, LNAI 4693, , KES 2007, Springer-Verlag, Germany, 2007.
- Apolloni, B., Howlett, R.J. and Jain, L.C. (Editors), Knowledge-Based Intelligent Information and Engineering Systems, Lecture Notes in Artificial Intelligence, Volume 3, LNAI 4694, KES 2007, Springer-Verlag, Germany, 2007.
- Nguyen, N.T., Grzech, A., Howlett, R.J. and Jain, L.C., Agents and Multi-Agents Systems: Technologies and Applications, Lecture Notes in artificial Intelligence, LNAI 4696, Springer-Verlag, Germany, 2007.

- Howlett, R.P., Gabrys, B. and Jain, L.C. (Editors), Knowledge-Based Intelligent Information and Engineering Systems, Lecture Notes in Artificial Intelligence, KES 2006, Springer-Verlag, Germany, LNAI 4251, 2006.
- Howlett, R.P., Gabrys, B. and Jain, L.C. (Editors), Knowledge-Based Intelligent Information and Engineering Systems, Lecture Notes in Artificial Intelligence, KES 2006, Springer-Verlag, Germany, LNAI 4252, 2006.
- Howlett, R.P., Gabrys, B. and Jain, L.C. (Editors), Knowledge-Based Intelligent Information and Engineering Systems, Lecture Notes in Artificial Intelligence, KES 2006, Springer-Verlag, Germany, LNAI 4253, 2006.
- Liao, B.-H., Pan, J.-S., Jain, L.C., Liao, M., Noda, H. and Ho, A.T.S., Intelligent Information Hiding and Multimedia Signal Processing, IEEE Computer Society Press, USA, 2007. ISBN: 0-7695-2994-1.
- Khosla, R., Howlett, R.P., and Jain, L.C. (Editors), Knowledge-Based Intelligent Information and Engineering Systems, Lecture Notes in Artificial Intelligence, KES 2005, Springer-Verlag, Germany, LNAI 3682, 2005.
- Skowron, A., Barthes, P., Jain, L.C., Sun, R., Mahoudeaux, P., Liu, J. and Zhong, N.(Editors), Proceedings of the 2005 IEEE/WIC/ACM International Conference on Intelligent Agent Technology, Compiegne, France, IEEE Computer Society Press, USA, 2005.
- Khosla, R., Howlett, R.P., and Jain, L.C. (Editors), Knowledge-Based Intelligent Information and Engineering Systems, Lecture Notes in Artificial Intelligence, KES 2005, Springer-Verlag, Germany, LNAI 3683, 2005.
- Khosla, R., Howlett, R.P., and Jain, L.C. (Editors), Knowledge-Based Intelligent Information and Engineering Systems, Lecture Notes in Artificial Intelligence, KES 2005, Springer-Verlag, Germany, LNAI 3684, 2005.
- Khosla, R., Howlett, R.P., and Jain, L.C. (Editors), Knowledge-Based Intelligent Information and Engineering Systems, Lecture Notes in Artificial Intelligence, KES 2005, Springer-Verlag, Germany, LNAI 3685, 2005.
- Negoita, M., Howlett, R.P., and Jain, L.C. (Editors), Knowledge-Based Intelligent Engineering Systems, KES 2004, Lecture Notes in Artificial Intelligence, LNAI. 3213, Springer, 2004
- Negoita, M., Howlett, R.P., and Jain, L.C. (Editors), Knowledge-Based Intelligent Engineering Systems, KES 2004, Lecture Notes in Artificial Intelligence, LNAI 3214, Springer, 2004
- Negoita, M., Howlett, R.P., and Jain, L.C. (Editors), Knowledge-Based Intelligent Engineering Systems, KES 2004, Lecture Notes in Artificial Intelligence, LNAI 3215, Springer, 2004
- Palade, V., Howlett, R.P., and Jain, L.C. (Editors), Knowledge-Based Intelligent Engineering Systems, Lecture Notes in Artificial Intelligence, LNAI 2773, Springer, 2003
- Palade, V., Howlett, R.P., and Jain, L.C. (Editors), Knowledge-Based Intelligent Engineering Systems, Lecture Notes in Artificial Intelligence, LNAI 2774, Springer, 2003
- Damiani, E., Howlett, R.P., Jain, L.C. and Ichalkaranje, N. (Editors), Proceedings of the Fifth International Conference on Knowledge-Based Intelligent Engineering Systems, **Volume 1**, IOS Press, The Netherlands, 2002.

- Damiani, E., Howlett, R.P., Jain, L.C. and Ichalkaranje, N. (Editors), Proceedings of the Fifth International Conference on Knowledge-Based Intelligent Engineering Systems, **Volume 2**, IOS Press, The Netherlands, 2002.
- Baba, N., Jain, L.C. and Howlett, R.P. (Editors), Proceedings of the Fifth International Conference on Knowledge-Based Intelligent Engineering Systems (KES'2001), **Volume 1,** IOS Press, The Netherlands, 2001.
- Baba, N., Jain, L.C. and Howlett, R.P. (Editors), Proceedings of the Fifth International Conference on Knowledge-Based Intelligent Engineering Systems (KES'2001), **Volume 2,** IOS Press, The Netherlands, 2001.
- Howlett, R.P. and Jain, L.C.(Editors), Proceedings of the Fourth International Conference on Knowledge-Based Intelligent Engineering Systems, IEEE Press, USA, 2000. **Volume 1.**
- Howlett, R.P. and Jain, L.C.(Editors), Proceedings of the Fourth International Conference on Knowledge-Based Intelligent Engineering Systems, IEEE Press, USA, 2000. **Volume 2**.
- Jain, L.C.(Editor), Proceedings of the Third International Conference on Knowledge-Based Intelligent Engineering Systems, IEEE Press, USA, 1999.
- Jain, L.C. and Jain, R.K. (Editors), Proceedings of the Second International Conference on Knowledge-Based Intelligent Engineering Systems, **Volume 1**, IEEE Press, USA, 1998.
- Jain, L.C. and Jain, R.K. (Editors), Proceedings of the Second International Conference on Knowledge-Based Intelligent Engineering Systems, **Volume 2**, IEEE Press, USA, 1998.
- Jain, L.C. and Jain, R.K. (Editors), Proceedings of the Second International Conference on Knowledge-Based Intelligent Engineering Systems, **Volume 3**, IEEE Press, USA, 1998.
- Jain, L.C. (Editor), Proceedings of the First International Conference on Knowledge-Based Intelligent Engineering Systems, **Volume 1**, IEEE Press, USA, 1997.
- Jain, L.C. (Editor), Proceedings of the First International Conference on Knowledge-Based Intelligent Engineering Systems, **Volume 2**, IEEE Press, USA, 1997.
- Narasimhan, V.L., and Jain, L.C. (Editors), The Proceedings of the Australian and New Zealand Conference on Intelligent Information Systems, IEEE Press, USA, 1996.
- Jain, L.C. (Editor), Electronic Technology Directions Towards 2000, ETD2000, **Volume 1**, IEEE Computer Society Press, USA, May 1995.
- Jain, L.C. (Editor), Electronic Technology Directions Towards 2000, ETD2000, **Volume 2**, IEEE Computer Society Press, USA, May 1995

## Book Series

Advanced Intelligence and Knowledge Processing, Springer-Verlag, Germany: www.springer.com/series/4738

- Zudilova-Seinstra, E. et al. (Editors), Trends in Interactive Visualization, Springer-Verlag, London, 2009.
- Monekosso, D., et al. (Editors), Intelligent Environments, Springer-Verlag, London, 2009.
- Chli, M. and de Wilde, P., Convergence and Knowledge Processing in Multi-agent Systems, Springer-Verlag, London, 2009
- Chein, M. and Mugnier, M.L., Graph-based Knowledge Representation, Springer-Verlag, London, 2009.
- Narahari, Y., et al., Game Theoretic Problems in Network Economics and Mechanism Design Solutions, Springer-Verlag, London, 2009.
- Zarri, G.P., Representation and Management of Narrative Information, Springer-Verlag, London, 2009.
- Hu, C., et al. (Editors), Knowledge Processing with Interval and Soft Computing, Springer-Verlag, London, 2008.
- Simovici, D.A. and Djeraba, C., Mathematical Tools for Data Mining, Springer-Verlag, London, 2008.
- Okada, A., et al. (Editors), Knowledge Cartography, Springer-Verlag, London, 2008.
- Nguyen, N.T., Advanced Methods for Inconsistent Knowledge Management, Springer-Verlag, London, 2008.
- Meisels, A., Distributed Search by Constrained Agents, Springer-Verlag, London, 2008.
- Camastra, F. and Vinciarelli, A., Machine Learning for audio, Image, and Video Analysis, Springer-Verlag, London, 2008.
- Kornai, A., Mathematical Linguistics, Springer-Verlag, London, 2008.
- Prokopenko, M. (Editor), Advances in Applied Self-Organising Systems, Springer-Verlag, London, 2008.
- Scharl, A., Environmental Online Communication, Springer-Verlag, London, 2007.
- Pierre, S. (Editor), E-Learning Networked Environments and Architectures, Springer-Verlag, London, 2007
- Karny, M. (Editor), Optimized Bayesian Dynamic Advising, Springer-Verlag, London, 2006.
- Liu, S. and Lin, Y., Grey Information: Theory and Practical Applications, Springer-Verlag, London, 2006.
- Maloof, M.A. (Editor), Machine Learning and Data Mining for Computer Security, Springer-Verlag, London, 2006.
- Wang, J.T.L., et al. (Editors), Data Mining in Bioinformatics, Springer-Verlag, London, 2005.
- Grana, M., et al. (Editors), Information Processing with Evolutionary Algorithms, Springer-Verlag, London, 2005.
- Fyfe, C., Hebbian Learning and Negative Feedback Networks,, Springer-Verlag, London, 2005.
- Chen-Burger, Y. and Robertson, D., Automatic Business Modelling,, Springer-Verlag, London, 2005.
- Husmeier, D., et.al. (Editors), Probabilistic Modelling in Bioinformatics and Medical Informatics, Springer-Verlag, London, 2005.

- Tan, K.C., et al., Multiobjective Evolutionary Algorithms and Applications, Springer-Verlag, London, 2005.
- Bandyopadhyay, S., et. al. (Editors), Advanced Methods for Knowledge Discovery from Complex Data, Springer-Verlag, London, 2005.
- Stuckenschmidt, H. and Harmelen, F.V., Information Sharing on the Semantic Web, Springer-Verlag, London, 2005.
- Abraham, A., Jain, L.C. and Goldberg, R., Evolutionary Multiobjective Optimization, Springer-Verlag, London, 2005.
- Gomez-Perez, et al., Ontological Engineering, Springer-Verlag, London, 2004.
- Zhang, S., et. al., Knowledge Discovery in Multiple Databases, Springer-Verlag, London, 2004.
- Ko, C.C., Creating Web-based Laboratories, Springer-Verlag, London, 2004.
- Mentzas, G., et al., Knowledge Asset Management, Springer-Verlag, London, 2003.
- Vazirgiannis, M., et al., Uncertainty Handling and Quality Assessment in Data Mining, Springer-Verlag, London, 2003.

## Advanced Information Processing, Springer-Verlag, Germany

- Harris, C., Hong, X. and Gan, Q., Adaptive Modelling, Estimation and Fusion from Data, Springer-Verlag, Germany, 2002.
- Ohsawa, Y. and McBurney, P. (Editors), Chance Discovery, Springer-Verlag, Germany, 2003.
- Deen, S.M. (Editor), Agent-Based Manufacturing, Springer-Verlag, Germany, 2003.
- Gasós J. and Hirsch B., e-Business Applications, Springer-Verlag, Germany, 2003.
- Chen, S.H. and Wang, P.P. (Editors), Computational Intelligence in Economics and Finance, Springer-Verlag, Germany, 2004.
- Liu, J. and Daneshmend, L., Spatial reasoning and Planning, Springer-Verlag, Germany, 2004.
- Wang, L. and Fu, X., Data Mining with Computational Intelligence, Springer-Verlag, 2005.
- Ishibuchi, H., Nakashima, T. and Nii, M., Classification and Modeling with Linguistic Information Granules, Springer-Verlag, Germany, 2005.

## Computational Intelligence and its Applications Series, IGI Publishing, USA

http://www.igi-pub.com/bookseries/details.asp?id=5

- Chen, S.-H., Jain, L.C. and Tai, C.-C. (Editors), Computational Economics: A Perspective from Computational Intelligence, IGI Publishing, 2006.
- Begg, R. and Palaniswami, M., Computational Intelligence for Movement Sciences, IGI Publishing, 2006.
- Fulcher, J. (Editor), Advances in Applied Intelligences, IGI Publishing, 2006.
- Zhang, D., et al., Biometric Image Discrimination technologies, IGI Publishing, 2006.

## Knowledge-Based Intelligent Engineering Systems Series, IOS Press, The Netherlands

http://www.kesinternational.org/bookseries.php

- Velasquez, J.D. and Palade, V., Adaptive Web Sites, IOS Press, Vol. 170, 2008.
- Zha, X.F. and Howlett, R.J., Integrated Intelligent System for Engineering Design, Vol. 149, IOS Press, 2006.
- Phillips-Wren, G. and Jain, L.C. (Editors), Intelligent Decision Support Systems in Agent-Mediated Environments, IOS Press, The Netherlands, Volume 115, 2005.
- Nakamatsu, K. and Abe, J.M., Advances in Logic Based Intelligent Systems, IOS Press, The Netherlands, Volume 132, 2005.
- Abraham, A., Koppen, M. and Franke, K. (Editors), Design and Applications of Hybrid Intelligent Systems, IOS Press, The Netherlands, Volume 104, 2003.
- Turchetti, C., Stochastic Models of Neural Networks, IOS Press, The Netherlands, Volume 102, 2004
- Wang, K., Intelligent Condition Monitoring and Diagnosis Systems, IOS Press, The Netherlands, Volume 93, 2003.
- Abraham, A., et al. (Editors), Soft Computing Systems, IOS Press, The Netherlands, Volume 87, 2002.
- Lee, R.S.T. and Liu, J.H.K., Invariant Object Recognition based on Elastic Graph Matching, IOS Press, The Netherlands, Volume 86, 2003.
- Loia, V. (Editor), Soft Computing Agents, IOS Press, The Netherlands, Volume 83, 2002.
- Motoda, H., Active Mining, IOS Press, The Netherlands, Volume 79. 2002.
- Namatame, A., et al. (Editors), Agent-Based Approaches in Economic and Social Complex Systems, IOS Press, The Netherlands, Volume 72, 2002.

## The CRC Press International Series on Computational Intelligence, The CRC Press, USA

http://www.crcpress.com/shopping_cart/products/product_series.asp?id=&series=747975&parent_id=&sku=1965&isbn=9780849319655&pc

- Teodorescu, H.N. Kandel, A. and Jain, L.C. (Editors), Intelligent systems and Technologies in Rehabilitation Engineering, CRC Press USA, 2001.
- Jain, L.C. and Fanelli, A.M. (Editors), Recent Advances in Artificial Neural Networks: Design and Applications, CRC Press, USA, 2000.
- Medsker, L., and Jain, L.C. (Editors) Recurrent Neural Networks: Design and Applications, CRC Press, USA, 2000.
- Jain, L.C., Halici, U., Hayashi, I., Lee, S.B. and Tsutsui, S. (Editors) Intelligent Biometric Techniques in Fingerprint and Face Recognition, CRC Press, USA, 2000.

- Jain, L.C. (Editor), Evolution of Engineering and Information Systems, CRC Press USA, 2000.
- Dumitrescu, D., Lazzerini, B., Jain, L.C. and Dumitrescu, A., Evolutionary Computation, CRC Press USA, 2000.
- Dumitrescu, D., Lazzerini, B., Jain, L.C., Fuzzy Sets and their Applications to Clustering and Training, CRC Press USA, 2000.
- Jain, L.C. and De Silva, C.W. (Editors), Intelligent Adaptive Control, CRC Press USA, 1999.
- Jain, L.C. and Martin, N.M. (Editors), Fusion of Neural Networks, Fuzzy Logic and Evolutionary Computing and their Applications, CRC Press USA, 1999.
- Jain, L.C. and Lazzerini, B., (Editors), Knowledge-Based Intelligent Techniques in Character Recognition, CRC Press USA, 1999.
- Teodorescu, H.N. Kandel, A. and Jain, L.C. (Editors), Soft Computing Techniques in Human Related Science, CRC Press USA, 1999.
- Jain, L.C. and Vemuri, R. (Editors), Industrial Applications of Neural Networks, CRC Press USA, 1998.
- Jain, L.C., Johnson, R.P., Takefuji, Y. and Zadeh, L.A. (Editors), Knowledge-Based Intelligent Techniques in Industry, CRC Press USA, 1998.
- Teodorescu, H.N., Kandel, A. and Jain, L.C. (Editors), Fuzzy and Neuro-fuzzy Systems in Medicine, CRC Press USA, 1998.

## International series on Natural and Artificial Intelligence, AKI

http://www.innoknowledge.com

- Apolloni, B., et al, Algorithmic Inference in Machine Learning, Advanced Knowledge International, Australia, 2006.
- Lee, R.S.T., Advanced Paradigms in Artificial Intelligence, Advanced Knowledge International, Australia, 2005.
- Katarzyniak, R, Ontologies and Soft Methods in Knowledge Management, Advanced Knowledge International, Australia, 2005.
- Abe, A. and Ohsawa, Y. (Editors), Readings in Chance Discovery, Advanced Knowledge International, Australia, 2005.
- Kesheng Wang, K., Applied Computational Intelligence in Intelligent Manufacturing Systems, Advanced Knowledge International, Australia, 2005.
- Murase, K., Jain, L.C., Sekiyama, K. and Asakura, T. (Editors), Proceedings of the Fourth International Symposium on Human and Artificial Intelligence Systems, University of Fukui, Japan, Advanced Knowledge International, Australia, 2004.
- Nguyen N.T., (Editor) Intelligent Technologies for inconsistent Knowledge Processing, Advanced Knowledge International, Australia, 2004.
- Andrysek, J., et al., (Editors), Multiple Participant Decision Making, Advanced Knowledge International, Australia, 2004.
- Matsuda, K., Personal Agent-Oriented Virtual Society, Advanced Knowledge International, Australia, 2004.

- Ichimura, T. and Yoshida, K. (Editors), Knowledge-Based Intelligent Systems for Healthcare, Advanced Knowledge International, Australia, 2004.
- Murase, K., and Asakura, T. (Editors), Dynamic Systems Approach for Embodiment and Sociality: From Ecological Psychology to Robotics, Advanced Knowledge International, Australia, 2003
- Jain, R., et al. (Editors), Innovations in Knowledge Engineering, Advanced Knowledge International, Australia, 2003.
- Graziella, T., Jain, L.C., Innovations in Decision Support Systems, Advanced Knowledge International, Australia, 2003.
- Galitsky, B., Natural Language Question Answering System Technique of Semantic Headers, Advanced Knowledge International, Australia, 2003.
- Guiasu, S., Relative Logic for Intelligence-Based Systems, Advanced Knowledge International, Australia, 2003.

## Series on Innovative Intelligence, World Scientific

http://www.worldscientific.com/

- Jain, L.C., Howlett, R.J., Ichalkaranje, N., and Tonfoni, G. (Editors), Virtual Environments for Teaching and Learning, World Scientific Publishing Company Singapore, Volume 1, 2002.
- Jain, L.C., Ichalkaranje, N. and Tonfoni, G. (Editors), Advances in Intelligent Systems for Defence, World Scientific Publishing Company Singapore, Volume 2, 2002.
- Howlett, R., Ichalkaranje, N., Jain, L.C. and Tonfoni, G. (Editors), Internet-Based Intelligent Information Processing, World Scientific Publishing Company Singapore, Volume 3, 2002.
- Zaknich, A., Neural Nets for Intelligent Signal Processing, World Scientific Publishing Company Singapore, Volume 4, 2003.
- Hirose, A., Complex Valued Neural Networks, World Scientific Publishing Company Singapore, Volume 5, 2003.
- Shapiro, A.F. and Jain, L.C. (Editors), Intelligent and Other Computational Intelligence Techniques in Insurance, World Scientific Publishing Company Singapore, Volume 6, 2003.
- Pan, J-S., Huang, H.-C. and Jain, L.C. (Editors), Intelligent Watermarking Techniques, World Scientific Publishing Company Singapore, Volume 7, 2004.
- Hasebrook, J. and Maurer, H.A., Learning Support Systems for Organisational Learning, World Scientific Publishing Company Singapore, Volume 8, 2004.

## Books

- Jain, L.C. and Nguyen, N.T. (Editors), Knowledge Processing and Decision Making in Agent-Based Systems, Springer-Verlag, Germany, 2009.
- Tolk, A. and Jain, L.C. (Editors), Complex Systems in Knowledge-based Environments, Springer-Verlag, Germany, 2009.

- Nguyen, N.T. and Jain, L.C. (Editors), Intelligent Agents in the Evolution of Web and Applications, Springer-Verlag, Germany, 2009.
- Jarvis, J., Ronnquist, R, Jarvis, D. and Jain, L.C., Holonic Execution: A BDI Approach, Springer-Verlag, 2008.
- Jain, L.C., Sato, M., Virvou, M., Tsihrintzis, G., Balas, V. and Abeynayake, C. (Editors), Computational Intelligence Paradigms: Volume 1 – Innovative Applications, Springer-Verlag, 2008.
- Phillips-Wren, G., Ichalkaranje, N. And Jain, L.C. (Editors), Intelligent Decision Making-An AI-Based Approach, Springer-Verlag, 2008.
- Fulcher, J. and Jain, L.C. (Editors), Computational Intelligence: A Compendium, Springer-Verlag, 2008.
- Sordo, M., Vaidya, S. and Jain, L.C.(Editors), Advanced Computational Intelligence Paradigms in **Healthcare 3**, Springer-Verlag, 2008.
- Virvou, M. And Jain, L.C.(Editors) , Intelligent Interactive Systems in Knowledge-Based Environments, Springer-Verlag, 2008.
- Sommerer, C., Jain, L.C. and Mignonneau, L. (Editors), The Art and Science of Interface and Interaction Design, Volume 1, Springer-Verlag, 2008.
- Nayak, R., Ichalkaranje, N. and Jain, L.C. (Editors), Evolution of the Web in Artificial Intelligence Environments, Springer-Verlag, 2008.
- Tsihrintzis, G. and Jain, L.C. (Editors), Multimedia Services in Intelligent Environments, Springer-Verlag, 2008.
- Tsihrintzis, G., Virvou, M., Howlett, R.J. and Jain, L.C. (Editors), New Directions in Intelligent Interactive Multimedia, Springer-Verlag, 2008.
- Holmes, D. and Jain, L.C. (Editors), Innovations in Bayesian Networks, Springer-Verlag, Germany, 2008.
- Magnenat-Thalmann, Jain, L.C. and Ichalkaranje, N., New Advances in Virtual Humans, Springer-Verlag, Germany, 2008.
- Jain, L.C., Palade, V. and Srinivasan, D. (Editors), Advances in Evolutionary Computing for System Design, Springer-Verlag, 2007.
- Baba, N., Handa, H. and Jain, L.C. (Editors), Advanced Intelligent Paradigms in Computer Games, Springer-Verlag, 2007.
- Chahl, J.S., Jain, L.C., Mizutani, A. and Sato-Ilic, M. (Editors), Innovations in Intelligent Machines 1, Springer-Verlag, 2007.
- Zharkova, V. and Jain, L.C. (Editors), Artificial Intelligence in Recognition and Classification of Astrophysical and Medical Images, Springer-Verlag, 2007.
- Pan, J-S., Huang, H.-C., Jain, L.C. and Fang, W.-C. (Editors), Intelligent Multimedia Data Hiding, Springer-Verlag, 2007.
- Yoshida, H., Jain, A., Ichalkaranje, A., Jain, L.C. and Ichalkaranje, N. (Editors), Advanced Computational Intelligence Paradigms in **Healthcare 1**, Springer-Verlag, 2007.
- Vaidya, S., Jain, L.C. and Yoshida, H. (Editors), Advanced Computational Intelligence Paradigms in **Healthcare 2**, Springer-Verlag, 2007.
- Jain, L.C, Tedman, R. and Tedman, D. (Editors), Evolution of Teaching and Learning in Intelligent Environment, Springer-Verlag, 2007.

- Sato, M. and Jain, L.C., Innovations in Fuzzy Clustering, Springer-Verlag, 2006.
- Patnaik, S., Jain, L.C., Tzafestas, S.G., Resconi, G. and Konar, A. (Editors), Innovations in Robot Mobility and Control, Springer-Verlag, 2006.
- Apolloni, B., Ghosh, A., Alpaslan, F., Jain, L.C. and Patnaik, S. (Editors), Machine Learning and Robot Perception, Springer-Verlag, 2006.
- Palade, V., Bocaniala, C.D. and Jain, L.C. (Editors), Computational Intelligence in Fault Diagnosis, Springer-Verlag, 2006.
- Holmes, D. and Jain, L.C. (Editors), Innovations in Machine Learning, Springer-Verlag, 2006.
- Ichalkaranje, N., Ichalkaranje, A. and Jain, L.C. (Editors), Intelligent Paradigms for Assistive and Preventive Healthcare, Springer-Verlag, 2006.
- Seiffert, U., Jain, L.C. and Schweizer, P. (Editors), Bioinformatics Using Computational Intelligence Paradigms, Springer-Verlag, 2005.
- Ghosh, A. and Jain, L.C. (Editors), Evolutionary Computation in Data Mining, Springer-Verlag, Germany, 2005.
- Phillips-Wren, G. and Jain, L.C.(Editors), Intelligent Decision Support Systems in Agent-Mediated Environments, IOS Press, The Netherlands, 2005.
- Silverman, B., Jain, A., Ichalkaranje, A. and Jain, L.C. (Editors), Intelligent Paradigms in Healthcare Enterprises, Springer-Verlag, Germany, 2005.
- Ghaoui, C., Jain, M., Bannore, V., and Jain, L.C. (Editors), Knowledge-Based Virtual Education, Springer-Verlag, Germany, 2005.
- Pal, N. and Jain, L.C.(Editors), Advanced Techniques in Knowledge Discovery and Data Mining, Springer-Verlag, London, 2005.
- Khosla, R., Ichalkaranje, N. and Jain, L.C.(Editors), Design of Intelligent Multi-Agent Systems, Springer-Verlag, Germany, 2005.
- Abraham, A., Jain, L.C. and van der Zwaag, B.(Editors), Innovations in Intelligent Systems, Springer-Verlag, Germany, 2004.
- Tonfoni, G. and Jain, L.C., Visualizing Document Processing, Mouton De Gruyter, Germany, 2004.
- Fulcher, J. and Jain, L.C.(Editors), Applied Intelligent Systems, Springer-Verlag, Germany, 2004.
- Damiani, E., Jain, L.C. and Madravio, M. (Editors), Soft Computing in Software Engineering, Springer-Verlag, Germany, 2004.
- Resconi, G. and Jain, L.C., Intelligent Agents: Theory and Applications, Springer-Verlag, Germany, 2004.
- Abraham, A., Jain, L.C. and Kacprzyk, J. (Editors), Recent Advances in Intelligent Paradigms and Applications, Springer-Verlag, Germany, 2003.
- Tonfoni, G. and Jain, L.C., The Art and Science of Documentation Management, Intellect, UK, 2003.
- Seiffert, U. and Jain, L.C. (Editors), Self-Organising Neural Networks, Springer-Verlag, Germany, 2002.
- Jain, L.C., Howlett, R.J., Ichalkaranje, N., and Tonfoni, G. (Editors), Virtual Environments for Teaching and Learning, World Scientific Publishing Company Singapore, 2002.

- Schmitt, M. Teodorescu, H.-N., Jain, A., Jain, A., Jain, S. and Jain, L.C. (Editors), Computational Intelligence Processing in Medical Diagnosis, Springer-Verlag, 2002.
- Jain, L.C. and Kacprzyk, J. (Editors), New Learning Paradigms in Soft Computing, Springer-Verlag, Germany, 2002.
- Jain, L.C., Chen, Z. and Ichalkaranje, N. (Editors), Intelligent Agents and Their Applications, Springer-Verlag, Germany, 2002.
- Jain, L.C. and De Wilde, P. (Editors), Practical Applications of Computational Intelligence Techniques, Kluwer Academic Publishers, USA, 2001.
- Howlett, R.J. and Jain, L.C. (Editors), Radial Basis Function Networks 1, Springer-Verlag, Germany, 2001.
- Howlett, R.J. and Jain, L.C. (Editors), Radial Basis Function Networks 2, Springer-Verlag, Germany, 2001.
- Teodorescu, H.N., Jain, L.C. and Kandel, A. (Editors), Hardware Implementation of Intelligent Systems, Springer-Verlag, Germany, 2001.
- Baba, N. and Jain, L.C., Computational Intelligence in Games, Springer- Verlag, 2001.
- Jain, L.C., Lazzerini, B. and Halici, U. (Editors), Innovations in ART Neural Networks, Springer-Verlag, Germany, 2000.
- Jain, A., Jain, A., Jain, S. and Jain, L.C. (Editors), Artificial Intelligence Techniques in Breast Cancer Diagnosis and Prognosis, World Scientific Publishing Company, Singapore, 2000.
- Jain, L.C. (Editor), Innovative Teaching and Learning in Intelligent Environment, Springer-Verlag, 2000.
- Jain, L.C. and Fukuda, T. (Editors), Soft Computing for Intelligent Robotic Systems, Springer-Verlag, Germany, 1998.
- Jain, L.C. (Editor), Soft Computing Techniques in Knowledge-Based Intelligent Engineering Systems, Springer-Verlag, Germany, 1997.
- Sato, M., Sato, S. and Jain, L.C., Fuzzy Clustering Models and Applications, Springer-Verlag, Germany, 1997.
- Jain, L.C. and Jain, R.K. (Editors), Hybrid Intelligent Engineering Systems, World Scientific Publishing Company, Singapore, 1997.
- Vonk, E., Jain, L.C. and Johnson, R.P., Automatic Generation of Neural Networks Architecture Using Evolutionary Computing, World Scientific Publishing Company, Singapore, 1997.
- Van Rooij, A.J.F., Jain, L.C. and Jain, L.C., Neural Network Training Using Genetic Algorithms, World Scientific Publishing Company, Singapore, December 1996
- Teodorescu, H.N., Watada, J. and Jain, L.C. (Editors), Intelligent Systems and Technologies, Springer-Verlag, Germany, 2009.
- Nicoletti, M.D.C. and Jain, L.C. (Editors), Computational Intelligence Techniques for Bioprocess Modelling, Supervision and Control, Springer-Verlag, Germany, 2009.
- Castellano, G.; Jain, L.C. and Fanelli, Anna Maria (Editors), Web Personalization in Intelligent Environments, Springer-Verlag, Germany, 2009.

- Pan, J.S., Huang, H.C. and Jain, L.C. (Editors), Information Hiding and Applications, Springer-Verlag, Germany, 2009.
- Damiani, E., Jeong, J., Howlett, R.J. and Jain, L.C. (Editors), New Directions in Intelligent Interactive Multimedia Systems and Services – 2, Springer-Verlag, Germany, 2009.
- Rakus-Andersson, E., Yager, R.R., Ichalkaranje, N. and Jain, L.C. (Editors), Recent Advances in Decision Making, Springer-Verlag, Germany, 2009.
- Wang, F.H., Pan, J.S. and Jain, L.C., Innovations in Digital Watermarking Techniques, Springer-Verlag, Germany, 2009.
- Finn, A.R. and Jain, L.C. (Editors), Innovation in Defence Support Systems, Springer-Verlag, Germany, 2009, under preparation.

## Book Chapters

- Jain, L.C., Lim, C.P. and Nguyen, N.T., Innovations in Knowledge Processing and Decision Making in Agent-Based Systems, Springer-Verlag, Germany, 2009, Chapter 1, pp. 1-18.
- Tweedale, J. and Jain, L.C., The Evolution of Intelligent Agents Within the World Wide Web, Springer-Verlag, Germany, 2009, Chapter 1, pp. 1-9.
- Tolk, A. and Jain, L.C., An Introduction to Complex Systems in Knowledge-based Environments, Volume 168, SCI Series, Springer-Verlag, Germany, 2009, pp. 1-5
- Pedrycz, W., Ichalkaranje, N., Phillips-Wren, G., and Jain, L.C., Introduction to Computational Intelligence for Decision Making, Springer-Verlag, 2008, pp. 75-93, Chapter 3.
- Tweedale, J., Ichalkaranje, N., Sioutis, C., Urlings, P. and Jain, L.C., Future Directions: Building a Decision Making Framework using Agent Teams, Springer-Verlag, 2008, pp. 381-402, Chapter 14.
- Virvou, M. and Jain, L.C., Intelligent Interactive Systems in Knowledge-Based Environments: An Introduction, Springer-Verlag, 2008, pp. 1-8, Chapter 1.
- Tsihrintzis, G. and Jain, L.C., An Introduction to Multimedia Services in Intelligent Environments, Springer-Verlag, pp. 1-10, 2008, Chapter 1.
- Jain, L.C. and Lim, C.P., An Introduction to Computational Intelligence Paradigms, Springer-Verlag, pp. 1-15, 2008 Chapter 1.
- Nayak, R. and Jain, L.C., An Introduction to the Evolution of the Web in an Artificial Intelligence Environment, pp. 1-15, 2008, Chapter 1.
- Nayak, R. and Jain, L.C., Innovations in Web Applications using Artificial Intelligence Paradigms, pp. 17-40, 2008, Chapter 2.
- Zharkova, V.V. and Jain, L.C., Introduction to Recognition and Classification in Medical and Astrophysical Images, Springer-Verlag, 2007, pp. 118, Chapter 1.
- Yoshida, H., Vaidya, S. and Jain, L.C., Introduction to Computational Intelligence in Healthcare, Springer-Verlag, 2007, pp. 1-4, Chapter 1.
- Huang, H.C., Pan, J.S., Fang, W.C. and Jain, L.C., An Introduction to Intelligent Multimedia Data Hiding, Springer-Verlag, 2007, pp. 1-10, Chapter 1.

- Jain, L.C., et al., Intelligent Machines :An Introduction, Springer-Verlag, 2007, pp. 1-9, Chapter 1.
- Jain, L.C., et al., Introduction to Evolutionary Computing in System Design, Springer-Verlag, 2007, pp. 1-9, Chapter 1.
- Jain, L.C., et al., Evolutionary Neuro-Fuzzy Systems and Applications, Springer-Verlag, 2007, pp. 11-45, Chapter 1.
- Do, Q.V, Lozo, P. and Jain, L.C., Vision-Based Autonomous Robot Navigation, in Innovations in Robot Mobility and Control, Springer-Verlag, 2006, pp. 65-103, Chapter 2.
- Tran, C., Abraham, A. and Jain, L., Soft Computing Paradigms and Regression Trees in Decision Support Systems, in Advances in Applied Artificial Intelligence, Idea Group Publishing, 2006, pp. 1-28, Chapter 1.
- Jarvis, B., Jarvis, D. and Jain, L., Teams in Multi-Agent Systems, in IFIP International Federation for Information Processing, Vol. 228, Intelligent Information Processing III, Springer, 2006, pp. 1-10, Chapter 1.
- Abraham, A. and Jain, L.C., Evolutionary Multiobjective Optimization, Springer-Verlag, 2005, pp. 1-6, Chapter 1.
- Sisman-Yilmaz, N.A., Alpaslan, F. and Jain, L.C., Fuzzy Multivariate Auto-Regression Method and its Application, in Applied Intelligent Systems, Springer, 2004, pp. 281-300.
- Lozo, P., Westmacott, J., Do, Q., Jain, L.C. and Wu, L., Selective Attention ART and Object Recognition, in Applied Intelligent Systems, Springer, 2004, pp. 301-320.
- Wang, F., Jain, L.C. and Pan, J., Genetic Watermarking on Spatial Domain, in Intelligent Watermarking Techniques, World Scientific, 2004, pp. 481-514, Chapter 17.
- Wang, F., Jain, L.C. and Pan, J., Watermark Embedding System based on Visual Cryptography, in Intelligent Watermarking Techniques, World Scientific, 2004, pp. 377-394, Chapter 13.
- Sioutis, C., Urlings, P., Tweedale, J., Ichalkaranje, N., Forming Human-Agent Teams within Hostile Environments, in Applied Intelligent Systems, Springer-Verlag, 2004, pp. 255-279.
- Jain, L.C. and Chen, Z., Industry, Artificial Intelligence In, in Encyclopedia of Information Systems, Elsevier Science, USA, 2003, pp. 583-597.
- Jain, L.C. and Konar, A., An Introduction to Computational Intelligence Paradigms, in Practical Applications of Computational Intelligence Techniques, Springer, 2001, pp. 1-38.
- Tedman, D. and Jain, L.C., An Introduction to Innovative Teaching and Learning, in Teaching and Learning, Springer, 2000, pp. 1-30, Chapter 1.
- Filippidis, A., Russo, M. and Jain, L.C., Novel Extension of ART2 in Surface Landmine Detection, Springer-Verlag, 2000, pp.1-25, Chapter 1.
- Jain, L.C. and Lazzerini, B., An Introduction to Handwritten Character and Word Recognition, in Knowledge-Based Intelligent Techniques in Character Recognition, CRC Press, 1999, 3-16.

- Filippidis, A., Jain, L.C. and Martin, N.N., "Computational Intelligence Techniques in Landmine Detection," in Computing with Words in Information/Intelligent Systems 2, Edited by Zadeh, L. and Kacprzyk, J., Springer-Verlag, Germany, 1999, pp. 586-609.
- Halici, U., Jain, L.C. and Erol, A., Introduction to Fingerprint Recognition, in Intelligent Biometric Techniques in Fingerprint and Face Recognition, CRC Press, 1999, pp.3-34.
- Teodorescu, H.N., Kandel, A. and Jain, L., Fuzzy Logic and Neuro-Fuzzy Systems in Medicine: A historical Perspective, in Fuzzy and Neuro-Fuzzy Systems in Medicine, CRC Press, 1999, pp. 3-16.
- Jain, L.C. and Vemuri, R., An Introduction to Intelligent Systems, in Hybrid Intelligent Engineering Systems, World Scientific, 1997, pp. 1-10, Chapter 1.
- Karr, C. and Jain, L.C., Genetic Learning in Fuzzy Control, in Hybrid Intelligent Engineering Systems, World Scientific, 1997, pp. 69-101, Chapter 4.
- Karr, C. and Jain, L.C., Cases in Geno- Fuzzy Control, in Hybrid Intelligent Engineering Systems, World Scientific, 1997, pp. 103-132, Chapter 5.
- Katayama, R., Kuwata, K. and Jain, L.C., Fusion Technology of Neuro, Fuzzy, GA and Chaos Theory and Applications, in Hybrid Intelligent Engineering Systems, World Scientific, 1997, pp. 167-186, Chapter 7.
- Jain, L.C., Medsker, L.R. and Carr, C., Knowledge-Based Intelligent Systems, in Soft Computing Techniques in Knowledge-Based Intelligent Systems, Springer-Verlag, 1997, pp. 3–14, Chapter 1.
- Babri, H., Chen, L., Saratchandran, P., Mital, D.P., Jain, R.K., Johnson, R.P. and Jain, L.C., Neural Networks Paradigms, in Soft Computing Techniques in Knowledge-Based Intelligent Systems, Springer-Verlag, 1997, pp. 15-43, Chapter 2.
- Jain, L.C., Tikk, D. and Koczy, L.T., Fuzzy Logic in Engineering, in Soft Computing Techniques in Knowledge-Based Intelligent Systems, Springer-Verlag, 1997, pp. 44-70, Chapter 3.
- Tanaka, T. and Jain, L.C., Analogue/Digital Circuit Representation for Design and Trouble Shooting in Intelligent Environment, in Soft Computing Techniques in Knowledge-Based Intelligent Systems, Springer-Verlag, 1997, pp. 227-258, Chapter 7.
- Jain, L.C., Hybrid Intelligent System Design Using Neural Network, Fuzzy Logic and Genetic Algorithms - Part I, Cognizant Communication Corporation USA, 1996, pp. 200-220, Chapter 9.
- Jain, L.C., Hybrid Intelligent System Applications in Engineering using Neural Network and Fuzzy Logic - Part II, Cognizant communication Corporation USA,1996, pp. 221-245, Chapter 10.
- Jain, L.C., Introduction to Knowledge-Based Systems, Electronic Technology Directions to the Year 2000, IEEE Computer Society Press USA, 1995, pp. 17-27, Chapter 1.
- Jain, L.C. and Allen, G.N., Introduction to Artificial Neural Networks, Electronic Technology Directions to the Year 2000, IEEE Computer Society Press USA, 1995, pp. 36-62, Chapter 2.

- Jain, L.C. and Karr, C.L., Introduction to Fuzzy Systems, Electronic Technology Directions to the Year 2000, IEEE Computer Society Press USA, 1995, pp. 93-103, Chapter 3.
- Jain, L.C. and Karr, C.L., Introduction to Evolutionary Computing Techniques, Electronic Technology Directions to the Year 2000, IEEE Computer Society Press USA, 1995, pp. 121-127, Chapter 4.
- Sato, M., Jain, L.C. and Takagi, H., Electronic Design and Automation, Electronic Technology Directions to the Year 2000, IEEE Computer Society Press USA, 1995, pp. 402-430, Chapter 9.
- Furuhashi, T., Takagi, H. and Jain, L.C., Intelligent Systems using Artificial Neural Networks, fuzzy Logic and Genetic Algorithms in Industry, Electronic Technology Directions to the Year 2000, IEEE Computer Society Press USA, 1995, pp. 485–4.
- Tweedale, J., Bollenbeck, F., Jain, L.C. and Urlings, P., Agent Transport Layer Adaptation System, Chapter 11, SCI 170, Springer-Verlag, Germany, pp. 247–273, 2009.
- Lim, C.P. and Jain, L.C., Advances in Intelligent Methodologies and Techniques, Chapter 1, Springer-Verlag, Germany, pp. 3-28, 2009.
- Nicoletti, M.D.C., Jain, L.C. and Giordano, R.C., Computational Intelligence Techniques as Tools for Bioprocess Modelling, Optimization, Supervision and Control, Chapter 1, Springer-Verlag, Germany, pp. 1-23, 2009.

# Author Index